SpringerBriefs in Economics

For further volumes:
http://www.springer.com/series/8876

Carol Yeh -Yun Lin · Leif Edvinsson
Jeffrey Chen · Tord Beding

National Intellectual Capital and the Financial Crisis in Brazil, Russia, India, China, Korea, and South Africa

Springer

Carol Yeh -Yun Lin
Department of Business Administration
National Chengchi University
Taipei, Taiwan

Jeffrey Chen
Accenture
Chicago, IL, USA

Leif Edvinsson
Universal Networking Intellectual Capital
Norrtälje, Sweden

Tord Beding
TC-Growth AB
Gothenburg, Sweden

ISSN 2191-5504 ISSN 2191-5512 (electronic)
ISBN 978-1-4614-6088-6 ISBN 978-1-4614-6089-3 (eBook)
DOI 10.1007/978-1-4614-6089-3
Springer New York Heidelberg Dordrecht London

Library of Congress Control Number: 2012951665

Printed on acid-free paper

Springer is part of Springer Science+Business Media (www.springer.com)

Foreword 1

The economic crisis is a consequence of many parallel factors which are all related to globalization and digitalization. My main concern, assessing this in more detail from the European perspective, is that the revolutionary global forces have not been taken early nor seriously enough by most national and regional decision makers. The Heads of European States and Governments have once again recalled the importance of fiscal consolidation, structural reform, and targeted investment to put Europe back on the path of smart, sustainable, and inclusive growth. The main question is how capable and ready are the national governments to tackling the complex and manifold issues of crises and to renewing even radically many of our public and private structures and processes.

The first basic requirement is that all the European Union Member States remain fully committed to taking the actions required at the national level to achieve the objectives of the Europe 2020 strategy. The second basic requirement is that the national and regional governments, as well as people, are ready for radical changes. This booklet and the other 11 booklets by the experienced authors focus on national intellectual capital and give necessary insights and facts to the readers and especially for readers in-depth systemic thinking of the interrelationships of NIC and economic recovery.

How should the national and regional decision makers tackle the existing knowledge of intangible capital? The focus needs to be more on the bottom-up approach stressing the developments on local and regional levels. I highlight our recent statements by the EU Committee of the Regions. The key priorities are to get more innovations out of research and to encourage mindset change towards open innovation.

The political decision makers are finally aware that the traditional indicators created for and used in industrial production cannot be applied to a knowledge-intensive, turbulent, and innovativeness-based global enterprise environment. Indicators that perceive the intangible dimensions of competitiveness—knowledge capital, innovation knowledge, and anticipation of the future—have been developed around the world, but their use has not yet become established in practice. This booklet accelerates the development and the use of these indicators.

This helps the local and regional, as well as central, governments in taking brave leaps forward on a practical level—giving greater ownership and involving all the stakeholders. This means the need of actions towards increasing the structural and relational capital of regions, both internally in communities of practice and in collaboration with others.

The new generation innovation activities are socially motivated, open, and collectively participated, complex and global by nature. The regions need to move towards open innovation, within a human-centered vision of partnerships between public and private sector actors, with universities playing a crucial role.

Regions should be encouraged to develop regional innovation platforms, which act as demand-based service centers and promote the use of international knowledge to implement the Europe 2020 strategy, smart specialization, and European partnerships according to the interests and needs of regions. For this to happen, we need to apply the new dynamic understanding of regional innovation ecosystems, in which companies, cities, and universities as well as other public and private sector actors (the "Triple Helix") learn to work together in new and creative ways to fully harness their innovative potential.

New innovative practices do not come about by themselves. One major potential is the use of public procurement. The renewing of the European wide rules must increase the strategic agility and activities of municipalities and other public operators as creators of new solutions. Especially the execution of pre-commercial procurement should be reinforced even more in combination with open innovation to speed up the green knowledge society development, i.e., for common reusable solutions in creating the infrastructures and services modern real-world innovation ecosystems are built upon. Conditions must be created that also allow for extensive development projects which address complex societal challenges and which take the form of risk-taking consortia.

One of our working instruments within the Committee of the Regions is the Europe 2020 Monitoring Platform, which broadly reviews and reflects the opinions and decisions on regional level all around Europe. It gives a flavor of cultural and other socioeconomic differences inside the EU. This brings an important perspective to the intellectual capital, namely the values and attitudes needed for citizens supporting policy makers on appropriate long-term investments and policies.

Emphasizing the importance of these issues, decision makers in all countries and regions worldwide need a deep and broad understanding of the critical success factors affecting the national intellectual capital. With all the facts and frames for thinking this booklet gives a valuable insight into today's challenges.

Markku Markkula
Advisor to the Aalto University Presidents
Member of the EU Committee of the Regions
Former Member of the Parliament of Finland

Foreword 2

Financial crisis—words very much heard today. What is all this about, actually, and how to get a grip on what we experience today? The booklet gives an important insight into the factors affecting competitiveness and productivity in modern knowledge society. We need to see behind the obvious, and we need to have increasingly "qualified guesses" as the character of the society and industry has fundamentally changed.

What is very important to notice is the shift towards intangible value creation beyond the deterministic phenomena we saw very clearly in the industrial era. Cost drivers were the important ones throughout the industry. Mass production, bigger is better, and very traditional productivity factor were the mantra.

However the production picture is changing. Increasingly value is created by the intangibles, often services related to the tangible components, and even totally in immaterial value creation, where perceptions and expectations determine the market value of the "extended product". We also see rapid change in organisational forms; we see new type of entrepreneurship growing besides the traditional industry clusters; we see smart specialisation of regions and countries.

This means also that there will be clearly different and complementary roles of the actors in innovation and value creation ecosystems. Large companies, small ones, even microenterprises together with the public sector are traditionally seen as the active partners in such innovation environments. The real issue in the dynamic markets is however that the end users are increasingly to be taken on board as *active subjects* for innovation and not merely treated as objects, customers. Markets need to be shaped and created in much more dynamic way than ever before. Open innovation beyond cross-licensing includes the societal capital as important intangible engine for productivity growth. Innovation happens only when the offering is meeting the demand. Otherwise we can only speak about inventions or ideas.

We need to have a close look at the intellectual capital and the different factors within it when we design our policy approaches. Short-term investments in process capital (infrastructures) and market capital seem to be very important for the manufacturing base as such, but at the same time measures for longer-term intellectual capital development and efficiency need to be taken.

Increasingly important is the *structure* and the open *processes* related to intangible capital and knowledge pools. For sustainable long-term development both the *human capital and the renewal capital are crucial*, as they are directly related to the innovation capability of the region. The correlation between these factors and the GDP growth is undisputable. In knowledge-intensive industries talent is attracting talent, and the connectivity which modern ICT provides makes this talent pool fluid across disciplines, organisations and geographical settings. It is imperative to modernise the innovation systems enabling the full dynamics needed for success in knowledge-intensive industries, beyond the traditional boundaries.

Measuring performance of innovation systems becomes increasingly complex due to the mash-up of different disciplines, having new types of actors and interactions between them. Hence the importance of analysis of the various components of the national intellectual capital (and equally on national innovation capability) as done in this booklet cannot be underestimated when making qualified guesses for operational choices to create functioning innovation ecosystems. The only predictable in true innovation is the unpredictability and the surprises. The role of the public sector is to drive strategy and measures enabling the unpredictable, and to catalyse a fluid, seamless and frictionless innovation system to grow, with strong interplay with the surrounding society.

We need to have courage to experiment, to prototype in real-world settings, to have all stakeholders involved to find and remove the friction points of innovation and to achieve sustainable innovation ecosystems for knowledge-intensive products and services.

I wish you interesting reading with this mind opening report.

Bror Salmelin
Advisor, Innovation Systems
European Commission
DG CONNECT

Foreword 3

The 2008 global financial crisis hit the whole world with unprecedented speed, causing widespread financial panic. Consumer confidence dropped to the lowest level since the Great Depression. Taiwan, with an export-dependent economy, was seriously impacted by the crisis and the unemployment rate hiked while household consumption levels dropped. At the onset of the financial crisis, Professor Lin was the Dean of Student Affairs here at National Chengchi University in Taipei, Taiwan. She was the dean in charge of financial aid and student loans and thus saw firsthand the direct impact the financial crisis had upon our students. The crisis was so devastating that Professor Lin, along with the university, was compelled to launch several new initiatives to raise money and help students weather the difficult times.

I am very glad that she took this painful experience to heart and set herself upon the task of investigating the impact of the crisis, trying to look into the causes and consequences for policy implications, not only for Taiwan but also for an array of 48 countries. In particular, she approaches the crisis from the perspective of "national intellectual capital," which is very important in today's knowledge-driven economy.

Taiwan is an example of a knowledge economy and has enjoyed the fame of being referred to as a "high-tech island." Without an abundance of natural resources, Taiwan's hardworking and highly educated population is the single most precious resource that the island has. Acknowledging the value of such human resources and intellectual capital, we established the Taiwan intellectual capital research center (TICRC) under my leadership in 2003. Ever since then, Taiwan's government has continuously funded the university to conduct relevant research projects aimed at enhancing the intellectual capital of Taiwan. Having been thus endowed with the responsibility of nourishing future leaders in the public and private sectors, we have focused on building up our strength in innovation, entrepreneurship, and technology management-related research and education.

To enhance intellectual capital research, we recently formed a joint team of professors for a four-year project in order to leverage their respective research capabilities. Through this project we hope to provide policy suggestions for the government by exploring the creativity, innovation, and intellectual capital at national, regional, city, and county levels. The goal is to come up with an intangible assets agenda for Taiwan's future sustainability. Professor Lin is an integral member in this research team.

Following her 2011 book *National Intellectual Capital: A Comparison of 40 Countries*, this booklet series is Professor Lin's second attempt at presenting her research, conducted under the sponsorship of TICRC, to international readers. As the Founding Director of TICRC and her President, I am honored to give a brief introduction of the value of this booklet series.

In comparison to her 2011 book, this series increased the number of countries studied to 48 and particularly focuses on the impact of intellectual capital on the 2008 global financial crisis. Rarely has an economic issue been systematically studied from the viewpoint of intangible assets, particularly at such a large scale of 48 countries. The research results show without a doubt that national intellectual capital is indeed an important economic development enhancer. In particular, the fact that countries with higher national intellectual capital experienced faster recoveries from the 2008 financial crisis provides a strong message for the policy makers.

In addition to providing insights into national policy, the booklet also summarizes the background of each country before the crisis, the key events during the crisis, economic development afterwards, and future prospects and challenges. Each volume affords readers a holistic picture of what happened in each country in an efficient manner. The linkage between national intellectual capital and this financial crisis also provides a different perspective of the crisis.

We are happy that Professor Lin continues to share her valuable research results with international readers. I sincerely hope that her insights can garner more attention concerning the benefits of developing national intellectual capital for the well-being of every nation.

Se-Hwa Wu
Professor, Graduate Institute of Technology
and Innovation Management
President, National Chengchi Univeristy, Taipei, Taiwan

Preface 1

There are "mounting risks of a breakup of the Euro zone." Such comments are frequent today on how the European leaders are handling the escalating crisis and its potential impact on non-European countries. But few leaders, reporters, or researchers are actually addressing the situation of national intellectual capital (NIC) and its signals. In addition to the financial crisis, is there an emerging NIC crisis as well? Why is it emerging? How should policy makers think about NIC? In what way does it need specific attention? When will the outcome and impact of taken NIC policy steps be realized?

In the midst of the European crisis, there are national interventions to address the issues mentioned above. In leading economical nations the investments going into intangibles now exceed tangibles and are positively correlated to income per capita. However, these still do not show up clearly in national mapping as well as policy-making insights. Therefore the New Club of Paris is focusing the knowledge agenda setting for countries on Societal Innovation (see www.new-club-of-paris.org).

Chairman Ben Bernanke of the U.S. Federal Reserve was addressing some of these same aspects in a key note speech in May 2011 hosted by Georgetown University: http://www.icapitaladvisors.com/2011/05/31/bernanke-on-intangible-capital/. OECD and the World Bank are developing NIC statistics, often based on the model from Corrado–Hultén. Japan has been developing both NIC and intangible assets (IA) at METI for some time now. Their research on IC/IA has resulted in a National IA Week with various key stakeholders, such as government agencies, universities, stock exchange, and enterprises. Japan is so far the only country in the world to hold such activities, and they have been doing so for the last 8 years. Australia, Singapore, South Korea, and China are currently undertaking various NIC initiatives. Other countries are also becoming more and more aware of NIC, with policy rhetoric centered on innovation, education, R&D, and trade. Despite this, the map for a more justified NIC navigation has been missing.

This booklet highlights NIC development for a number of countries, based on 28 different indicators, aggregated into four major NIC components of human capital, market capital, process capital, and renewal capital. The model here is a refined and verified statistical model in comparison to the Corrado–Hultén model. We call it the

L-E-S model after the contributors Lin–Edvinsson–Stahle. Based on a deeper understanding and the timeline pattern it sets forth, this model will add to a better NIC navigation, not to mention knowledge agenda setting for countries.

Upon looking at a global cluster NIC map, it is evident that the top leading countries seem to be small countries, especially Singapore, the Nordic countries, Hong Kong, and Taiwan. For the USA, Finland, and Sweden around 50 % or more of its economical growth is related to NIC aspects. Sweden, Finland, Switzerland, the USA, Israel, and Denmark are strongly influenced in its GDP growth by focusing on Renewal Capital.

It might be that we will see a clearer map of the NIC ecosystem and drivers forwealth emerge in the extension of this ongoing unique research of NIC. This booklet will present a NIC map for various clusters of countries. It can be used for bench marking as well as bench learning for policy prototyping. The starting point is awareness and thinking of NIC and its drivers for economic results. Based on this more refined navigation, NIC metrics can be presented.

Deeper understanding will emerge from this research, such as the scaling up of limited skilled human capital in one nation by using the globalized broadband technologies for migration and flow of knowledge (such as tele-medicine or mobile banking in Africa). This is also referred to as the IC multiplier. It might also be the way the old British Commonwealth was constructed, but without the IC taxonomy. In modern taxonomy it might be the shaping of NIC alliances for the migration and flow of IC between nations?

Another understanding that might emerge for policy making is the issue of employment versus unemployment. The critical understanding will be deployment of IC drivers. This will require another networked workforce of value networkers on a global scale, such as volunteering software and apps developers. However such volunteers do not show up in traditional statistics, for the mapping on behalf of policy makers.

On another level there might be clear gap analyses between nations to support the vision process of a nation. On a deeper level it is also a leadership responsibility to address the gap of NIC positions versus potential positions. Such a gap is in fact a liability to the citizens to be addressed in due time.

This will take us to the need for the continuous renewal of social systems. The so-called Arab Spring is explained by some as resulting from three drivers: lack of renewal of social systems, Internet, and soccer as cross class interaction space. The lack of social renewal and innovation is most likely critical early warning signals. For Greece, we can see such a tipping point occurred back in 1999.

On a global scale we might see that the concern for the Euro zone crisis should and can be explained by a deeper and supplementary understanding of national intellectual capital, in addition to financial capital. So we need to refine our NIC understanding, NIC mapping, NIC metrics, and NIC organizational constructs into societal innovation for the benefit of wealth creation of subsequent generations.

Leif Edvinsson
The World´s First Professor of Intellectual Capital
Chairman and Co-founder of New Club of Paris

Preface 2

Our first book *National Intellectual Capital: A Comparison of 40 Countries* was published in early 2011, at a time when the 2008 global financial crisis had been declared over yet the European region was still plagued with sovereign debt problems. Before we finalized the book, we were able to retrieve some of our raw data concerning the troubled countries, such as Greece, Iceland, Ireland, Portugal, and Spain. The results of our analysis based on the data spanning 1995–2008 revealed some early warning signs of the financial turmoil in those countries. In my preface of that book, I mentioned the warning signs might reveal only the tip of an iceberg. At that time, my coauthor, Professor Edvinsson and I decided to do a follow-up study to trace the development of national intellectual capital (NIC) in as many countries as possible, particularly through the lens of the 2008 global financial crisis. This 12 booklet series is the result of that determination.

The 2008 global financial crisis came with unexpected speed and had such a widespread effect that surprised many countries far from the epicenter of the initial U.S. sub-prime financial problem, geographically and financially. According to reports, no country was immune from the impact of this financial crisis. Such development clearly signifies how closely connected the world has become and the importance of having a global interdependent view. By reporting what happened during 2005–2010 in 48 major countries throughout the world, this booklet series serves the purpose of uncovering national problems before the crisis, government coping strategies, stimulus plans, potential prospects and challenges of each individual country, and the interdependence between countries. The 6 years of data allow us to compare NIC and economic development crossing before, during, and after the financial crisis. They are handy booklets for readers to have a quick yet overall view of countries of personal interest. The list of 48 countries in 11 clusters is provided in the appendix of each booklet.

Searching for financial crisis-related literature for 48 countries is itself a very daunting task, not to mention summarizing and analyzing it. For financial crisis-related literature, we mainly relied on the reports and statistics of certain world organizations, including OECD, World Bank, United Nations, international monetary fund (IMF), European Commission Office, the U.S. Congressional Research Service, the

U.S. Central Intelligence Agency, and international labor office (ILO). Some reliable research centers, such as the National Bureau of Economic Research in the U.S., World Economic Forum, the Heritage Foundation in the U.S., and government websites from each country were also our sources of information. Due to the requirement of more update and comprehensive information, we were not able to use as much academic literature as we would have liked, because it generally covers a very specific topic with time lag and with research methods not easily comprehended by the general public. Therefore, we had to resort to some online news reports for more current information.

In the middle of 2012, the lasting financial troubles caused the European economy to tilt back into a recession, which also slowed down the economic growth across the globe. However almost 4 years have passed since the outbreak of the global financial crisis in late 2008; it is about time to reflect on what happened and the impact of the financial crisis. By comparing so many countries, we came to a preliminary conclusion that countries with faster recovery from the financial crisis have higher national intellectual capital than those with slower recovery. In other words, countries that rebounded fast from the crisis generally have solid NIC fundamentals, including human capital, market capital, process capital, and renewal capital. We also found that the higher the NIC, the higher the GDP per capita (ppp). This booklet series provides a different perspective to look beyond the traditional economic indicators for national development.

In an era when intangible assets have become a key competitive advantage, investing in national intellectual capital development is investing in future national development and well-being.

Enjoy!

Carol Yeh-Yun Lin
Professor, Dept. of Business Administration
National Chengchi University, Taiwan
Taiwan Intellectual Capital Research Center (TICRC)

Executive Summary

Human capital and renewal capital will be the key intangible assets that decide the forerunners for sustainable national development among emerging countries.

How can national intellectual capital (NIC) be supportive of national development and sustainable national well-being? What are the key factors that facilitate emerging countries in becoming developed countries in the future? What will be the differentiating factor for current emerging countries 20 years from now?

One main cause of the 2008 global financial crisis was the failure of conventional financial metrics and accounting systems to detect potential risks due to non-transparent information disclosure. Our earlier NIC research has revealed certain warning signs of impending financial crisis for Greece, Iceland, and Ireland. Such findings indicate that NIC, despite being intangible, can provide valuable insights into future risk control and strategy formulation. This booklet looks into the connections between the financial crisis and NIC development for Brazil, Russia, India, China, Korea, and South Africa (*BRICKS*).

Research analysis of intangible NIC presented in various figures and tables in this booklet largely aligns with the development of tangible economic indicators, such as GDP growth. Data of 48 countries indicate that the higher the NIC index, the higher the GDP per capita (ppp), accentuating the value of NIC in major countries throughout the world. **For the 6-year average of NIC ranking,** *Brazil ranks #42, Russia #38, India #44, China #40, Korea #22, and South Africa #37.*

The 2008 financial crisis came with an unexpected speed and spread into a global economic shock, resulting in a number of bank failures in western countries as well as political and societal leadership challenges. During this period, economies worldwide slowed, credits tightened, and international trade declined. Governments and central banks worldwide responded to the crisis with unprecedented fiscal stimuli, monetary policy expansions, and institutional bailouts in their respective countries. Although the global financial crisis was declared over by the end of 2009, European economies were tilted back into recession again in the second half of 2011 due to lasting sovereign debt problems.

All six BRICKS countries depend heavily on exports and were hard hit by the crisis, with Russia having the greatest drop (7.88 % GDP per capita (ppp)) in 2009 due to its single energy industry export. However, all the countries rebounded to positive growth in 2010. The general government debt level is relatively low and stable in BRICKS countries when compared to that of European nations. Over that 6-year period, Russia and China had about 10 % and 20 % of GDP government debt, respectively; Korea and South Africa about 30 %; and Brazil and India around 50 %. Possessing good pre-crisis surpluses, BRICKS' expanding stimulus spending during the financial crisis did not affect their debt level too much.

During the financial crisis, unemployment rate was a key position indicator of how devastating the situation was. Over the 6 years, Brazil, Russia, and South Africa have reduced their unemployment rate; China and Korea maintained a similar level of unemployment, and India is the only country that has a continuous unemployment rise after the financial crisis. For South Africa, although its rate decreased from 2005 to 2010, its 24 % unemployment rate for 2010 is still shockingly high and poses a great threat for its future development. With regard to consumer price inflation, although high inflation has troubled most of these countries (except Korea) for years, each country has been able to suppress inflation after the financial crisis with the exception of India. India ended up with 10.5 % inflation in 2010, which was more than double its 2005 level.

NIC ranking of the BRICKS countries fell in the last quartile (except Korea) among the 48 countries, indicating a huge "intellectual capital in waiting." Russia stands out particularly in the area of renewal capital and India has the best market capital among the six countries. Unlike its fast economic growth, China is losing its market capital, process capital, and NIC international competitiveness with a large-scale ranking drop over the 6 years. On the contrary, Brazil has large-scale ranking gains in market capital and renewal capital.

The NIC 3D trajectory analysis detects the enhancing and impeding factors of GDP per capita (ppp) growth. It turns out that government-related and R&D-related issues are the key contributing factors and human capital-related issues are the main inhibiting factors of GDP growth in the BRICKS countries. Using Japan's GDP as a benchmark, India has the longest distance to cover, followed by South Africa, Brazil, Russia, China, and Korea.

This economic crisis provides an ideal opportunity for nations to examine the soundness of their economic system and the effectiveness of national governance related to NIC. The following policy implications are drawn from our research findings. Readers can refer to Chap. 5 for the rationale behind these implications.

1. National intellectual capital development goes together with the economic development and should be regarded as an enhancer of national sustainable growth.
2. The BRICKS countries are under subtle co-opetition (cooperation and competition) pressure and managing the relationship constructively will lead to a progressing world.
3. Striking a balance between export encouragement and domestic markets expansion will build more resilient economies for BRICKS countries.

4. Human capital and renewal capital will be the key intangible assets that decide the forerunners for sustainable national development among emerging countries.
5. Nations should not be blinded by minor internal economic changes; identifying one's own position on the global landscape and taking proper coping measures are crucial for a nation's sustainable development.
6. Developing intangible national features helps accumulate better market capital and process capital.
7. Different from a majority of positive economic statistics about India, NIC analysis reveals warning signals for the country.
8. Russia should continue its industry diversification and develop its small and medium enterprises.
9. Korea is an ideal benchmark for showing the transformation from an emerging country to a newly developed country.

The original BRIC and later BRICS (adding South Africa) countries have caught worldwide attention, mainly due to their economic performance and future prospect. In general, they have weathered the financial crisis relatively well and are expected to lift global economy in the future with their abundant natural and human resources. Korea's balanced NIC, strong market capital, and process capital development even during the financial crisis set a good example of how to transform from an emerging country into a newly developed country over a relatively short period.

As of mid-2012, the world economy has been slowed down by the pending debt problems in the Euro zone. The BRICKS economies were also affected and exhibited slower growth. Each country's resilience to crisis will again be tested if the world plunges back into a recession.

Our NIC intelligence suggests that in an era where the intangible asset has become a key competitive and sustainable advantage, investing in national intellectual capital development is equivalent to investing in future economic development. National intellectual capital evolution can be enhanced from both a local culture standpoint and a global interconnectivity standpoint by social media.

Based on the emerging new insights of values, societal history as well as citizen relationships, a key focus for the future will be on the fusion of national intellectual capital, social service innovation, and societal innovation, for the enabling of a new societal fabric.

Contents

List of Figures

List of Tables

Appendices

Chapter 1
Introduction

In 2012, aftermath conditions of the 2008 global financial crisis vary from country to country. Some have recovered from the turmoil while others are still struggling to stand on their own financial feet. In hindsight, it is valuable to reflect what had happened during the last few years in different countries to gain some insights for future implications.

What started off as sub-prime mortgage problems in the financial sector of advanced economies snowballed into the deepest and most widespread financial and economic crisis in the past 80 years. With almost synchronized recessions worldwide, global GDP contracted for the first time since World War II according to the International Monetary Fund (IMF). Governments and central banks around the world responded to the crisis with unprecedented stimulus packages in order to maintain financial order and help industries, private companies, and citizens wade through the difficulties.

Initially, the collapse of sub-prime bonds in the U.S. resulted in shortage in the global money market, huge amounts of foreign direct investments (FDI) were withdrawn all over the world to meet the liquidity needs in their home countries. Owing to very negative cash flow conditions, many companies slashed their production levels and canceled or postponed investment projects (Mendonça 2010). With the tight money market, export demands dropped and unemployment increased. Adding to the scene, some European Union countries had been overlending and overspending throughout the last decade due to easy EU money with low interest rates. Consequently, banks in many countries did not have enough money to support the repayment of foreign debts and corporate loans. As such, confidence fell, which directly hit the consumer market as the global financial crisis set in.

At the early stage of the crisis, management scholars criticized the inability of traditional accounting systems to reveal intangible assets that explain hidden values as well as risks for proper decision making (Reavis 2009). In line with such criticism, intellectual capital (Edvinsson and Malone 1997) advocates the values of intangible human capital, social capital and the like and has gained increasing attention in today's knowledge-driven economy.

C.Y.-Y. Lin et al., *National Intellectual Capital and the Financial Crisis in Brazil, Russia, India, China, Korea, and South Africa*, SpringerBriefs in Economics 18, DOI 10.1007/978-1-4614-6089-3_1, © The Author(s) 2013

It is our deep belief that national intellectual capital, albeit intangible, can provide valuable insight to policy makers regarding future risk control and strategy formulation. Our previous book, *National Intellectual Capital: A Comparison of 40 Countries* (Lin and Edvinsson 2011; www.nic40.org), was born out of this belief and traces the national intellectual capital development of 40 countries over 14 years (1995–2008). The data analysis revealed certain warning signs of impending financial crisis for countries such as Greece, Iceland, and Ireland (Lin and Edvinsson 2011: 327–333). As a follow-up study, this booklet series is an attempt to further explore the connections between the financial crisis and the national intellectual capital (NIC) development.

The booklet series, in its entirety, will examine the national intellectual capital statuses of 48 countries from the period of 2005–2010 to glean new understanding about whether there is a NIC development pattern that distinguishes fast recovery nations from slow recovery ones. This is presented through a series of 11 country clusters, with one booklet focusing on one particular cluster. The clusters are decided based upon several factors: geographical proximity, geographical size, or the country's phase of economic development. Focusing on one cluster at a time, we first probe the areas of concern within a single country and then compare them between multiple countries to see whether the situation before and after the crisis can be explained by the intangible NIC. Our data comes from the well-recognized International Institute for Management Development (IMD) in Switzerland. The IMD has been publishing yearly rankings of World Competitiveness for around two decades. It is our hope that the insight in this booklet series can provide a different perspective of the financial crisis for future policy implications.

This is the third volume and will focus on the six BRICKS countries, comprised of Brazil, the Russian Federation (referred as Russia hereafter), India, China, South Korea (referred as Korea hereafter), and South Africa. The term "BRIC" was first coined by Goldman Sachs in its 2003 report, which speculated that by 2050 Brazil, Russia, India, and China would be wealthier than most of the current major economic powers. The BRIC concept posits that China and India will become the world's dominant suppliers of manufactured goods and services, while Brazil and Russia will become similarly dominant as suppliers of raw materials. Adding Korea to the acronym is modeling Dahlman's (2009) categorization from the viewpoint that Korea moved rapidly from a very poor low-income country in the 1960s to a high-income economy by 1996 when it joined the OECD. Such characteristics can be the role model of the BRIC countries, although Korea has a much smaller territory. In addition, the original BRIC becomes BRICS, with South Africa officially accepted as a full member of the group in late December 2010 (Mu 2010). As a result, this volume adds two countries—Korea and South Africa—to the original acronym to make the six-country "BRICKS."

This booklet will first provide an economic background of these six countries as a whole before going into each individual economy's development. Through this process, the authors hope to paint a general picture of economic conditions and provide a basis for our dataset and analysis in future sections.

Since the economic history of these six countries goes back centuries, it is impossible to cover the entire spectrum in our work. As such, this background discussion will consider events in the relatively recent history from 2005 onwards that have the most direct impact upon the current economic conditions of each specific country. In addition, particular attention will be given to the 2008 global financial crisis. In doing so, the authors hope that the background, in conjunction with our later data and analysis, will provide a "before, during, and after" picture of what happened from a macroeconomic and intangible assets viewpoint.

Chapter 2 will briefly introduce the impact of 2008 financial crisis on the six countries. Chapter 3 discusses the NIC development. Chapter 4 describes issues beyond the financial crisis and Chapter 5 concludes with future perspectives and policy implications.

Economic Background

This section briefly introduces the characteristics of the first five countries—BRICK (Dahlman 2009) and then the membership of South Africa. Brazil has been a commodity exporter in the global economy, starting from the 1990s and still remains a leading exporter in the world. Russia is a former superpower nation that saw its economy contracted with its gradual transition into a free market economy. India's recent rapid growth is being powered by knowledge-intensive service exports facilitated by information technology. China is a manufacturing powerhouse that has moved towards a market economy by integrating with the global economy and efficiently tapping global knowledge. Korea is a new member of high-income nation and moved rapidly from a labor-intensive manufacturing to medium- and high-technology products exporter, especially with little reliance on FDI.

South Africa sought BRIC membership in 2010 and was officially accepted as a full member on December 24, 2010. Geopolitically, South Africa's membership gives BRICS a four-continent breadth, influence, and trade opportunities (Smith 2011). Commercially, this powerful bloc of emerging economies is expected to increase inter-trade amongst each other to the range of US$400 to $500 billion by the year 2015 (SAinfo 2011). According to the IMF, the total population of almost three billion within all the BRICS countries in 2012 will account for as much as 61% of global growth in three years' time. Furthermore, what South Africa brings into the bloc is not only the single nation of South Africa, but rather the larger African market of roughly a billion people (SAinfo 2011).

After the financial crisis, these six countries were widely known for their fast recoveries. To paint a general picture of their global competitiveness in the ex-ante and ex-post years of the financial crisis, we introduce hereunder the Global Competitiveness Index (GCI) published by the World Economic Forum for the reader's reference. This index is robust as it takes into account 12 distinct pillars[1]

[1] The 12 pillars include: institutions, infrastructure, macroeconomic environment, health and primary education, higher education and training, goods market efficiency, labor market efficiency, financial market development, technological readiness, market size, business sophistication, and innovation.

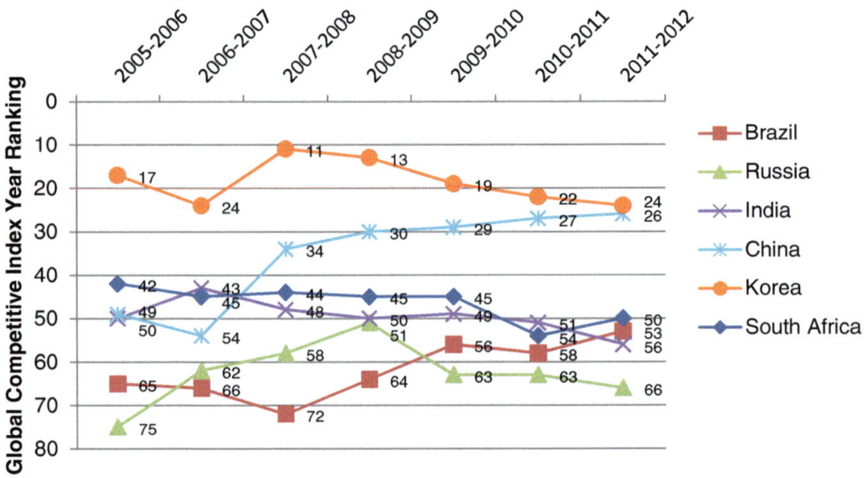

Fig. 1.1 GCI ranking of the six BRICKS countries

containing areas such as basic requirements, efficiency enhancers, and innovation factors that contribute to a nation's overall economic strength. Based on the commonly accepted economic theories, the development of a total of 142 countries was split into three stages in which different factors play dominant roles in determining the outcome of a country's economy. Stage 1, Stage 2, and Stage 3 are, respectively, characterized by being factor, efficiency, and innovation driven (Schwab 2011). As of this publication, India is categorized as Stage 1, China and South Africa Stage 2, Brazil and Russia Stage 2.5, and Korea Stage 3.

Plotting each nation's annual ranking against a time series of seven periods, Fig. 1.1 displays a rough pictorial overview of the six countries' global competitiveness before and after the financial crisis. As an indicator of economic strength in 2011–2012, it can be seen that Brazil #53, Russia #66, India #56, China #26, Korea #24, and South Africa #50 rank in two groups. China and Korea are around the first quartile and the other four countries are within the third quartile. Particularly, China as a Stage 2 country is ranked ahead of Brazil and Russia (both stage 2.5 countries) in the GCI index. In addition, China is rapidly catching up to South Korea as well. In contrast to China's strong upward trend, three countries are currently losing global competitiveness according to the graph. India dropped from 50 to 56, Korea from 17 to 24, and South Africa from 42 to 50. China has the fastest progression in global competitiveness in this group.

In September 2011, OECD reported that major economies were tilting back into recession as sovereign debt problems continued to ail the Euro zone (Bryant 2011). In Brazil, the economy picked up in 2012, driven by private consumption and investment. However, OECD (2012a) suggested that Brazil's weak manufacturing performance be addressed by further reducing the tax burden and tax complexity, deepening long-maturity financial markets and improving infrastructure.

2012 growth in emerging Asia is predicted at 7–7.5%, which is anchored by China with its growth rate of more than 8%. For India, close to 7% growth is also forecasted in 2012. Korean output growth, following a slowdown in late 2011, is projected to pick up in 2012, led by a rebound in exports and domestic demand. This development will help Korea achieve an output growth of 4% in 2013 (OECD 2012b). For overall Asia and the Pacific region, growth is projected at 6% in 2012 and about 6.5% in 2013 (IMF 2012a). Stronger economic and policy fundamentals have helped buffer the region's economies by limiting adverse financial woes spilled by European banks. In a press briefing, IMF (Singh et al. 2012) reported that leading indicators for Asia have strengthened, growth is gaining momentum, and domestic demand in many Asian countries remained strong.

For Russia, higher oil prices and the easing of Euro area tensions will allow its economy to continue to grow at above 4% in 2012 and 2013; its budget is projected to remain in surplus in 2012–2013 (OECD 2012c). In 2011, South Africa's growth slowed due to weakness in its major European trading partners (IMF 2012b). To mitigate South Africa's widening current account deficit and help contain inflation pressures, OECD (2012d) suggested its fiscal consolidation be stepped up. In addition, the still extremely high unemployment rate remains urgent.

The next chapter will give a brief background and qualitative analysis of the financial crisis as it relates to these six economies.

Chapter 2
Impact of the 2008 Global Financial Crisis

In order to present the impact of the 2008 global financial crisis, this chapter will graphically compare the real GDP growth, total general government debt, unemployment rate, and consumer price inflation of the six countries from 2005 to 2010. Then, the chapter will elaborate qualitatively on the impact of the financial crisis on each country's economy individually in the sequence of *B*razil, *R*ussia, *I*ndia, *C*hina, *K*orea, and *S*outh Africa (BRICKS).

BRICKS countries, except Korea, are emerging economies with sizeable domestic markets. For the original four BRIC countries, strong economic growth since 2001 has increased disposable incomes, thereby creating a large stratum of middle class consumers. In 2002, they had 20.6 million households with an annual disposable income over US$10,000. By 2007 this number catapulted to 90.1 million households. In 2007, consumer expenditure as a share of GDP amounted to 35.0% in China, 48.0% in Russia, 54.1% in India, and 61.0% in Brazil (Eghbal 2008). After a decade of growth, BRIC economies have built up strong consumer demand, which is the primary engine for growth. Although South Africa with an area 12 times as large as South Korea, its population of only 48 million is similar to South Korea. However, as mentioned earlier, South Africa has a market of the whole Africa with around one billion people for it to tackle.

At the initial stage of the 2008 global financial crisis, there was a general belief in these countries that they could remain largely insulated from the crisis and provide an alternative engine of growth for the world economy. As the financial crisis continued to roil credit and stock markets around the globe, it seemed that no country was being spared the consequences; the BRICKS countries were no exception (Khemka et al. 2008; Draper et al. 2009; Bajpai 2010a, b; Kim 2010c).

C.Y.-Y. Lin et al., *National Intellectual Capital and the Financial Crisis in Brazil, Russia, India, China, Korea, and South Africa*, SpringerBriefs in Economics 18,
DOI 10.1007/978-1-4614-6089-3_2,

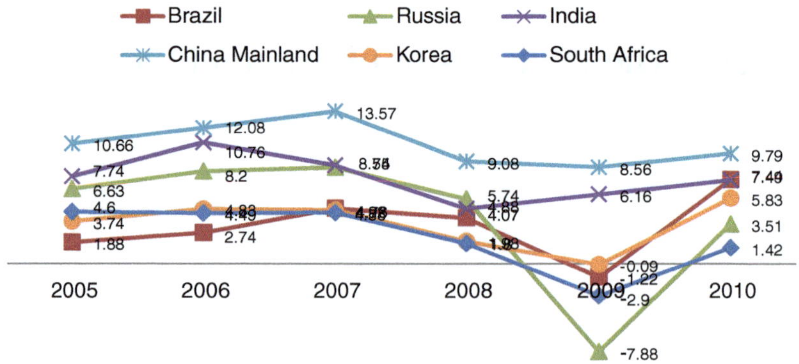

Fig. 2.1 Real GDP growth per capita of Brazil, Russia, India, China, Korea, and South Africa from 2005 to 2010

Comparisons of the Six Countries

This section presents four graphs in order to examine the BRICKS countries as a whole; namely, the percentage of real GDP growth per capita, total general government debt percentage of GDP, unemployment rate of labor force, and consumer price inflation from 2005 to 2010.

Figure 2.1 shows that from 2007 to 2008, all the countries had GDP growth decline, except Brazil remained relatively stable. In 2009, China and India are the only two countries that still had strong growth, with 8.56% vs. 6.16%, respectively. The other four countries in this cluster all had an obvious dip in real GDP growth in 2009, with Russia declining the most to −7.88%. However, all six economies rebounded with positive growth in 2010. Russia has the largest scale of GDP growth fluctuation over the six years and ended in 2010 surpassing South Africa only.

Figure 2.2 shows the total general government debt percentage of GDP. The reason for reporting government debt is based on past research findings. After researching 800 years financial crises, Reinhart and Rogoff (2009) commented that the overhang of public and private debt is the most important impediment to a normal recovery from recession. As such, we have included government debt as an indicator of the health of the recovery. Figure 2.2 indicates that the government debt percentage of GDP for the BRICKS countries remains relatively stable over the six years, with roughly three clusters.

Brazil and India form the high-debt cluster of around 50% GDP; Korea and South Africa form the middle-debt cluster of around 30% GDP; and China and Russia make up the low-debt cluster of below 20% GDP. In 2005, the debt level of China (17.64%) and Russia (14.16%) were relatively close. With the oil price hike, Russia as a major oil export country continued to reduce its government debt to only

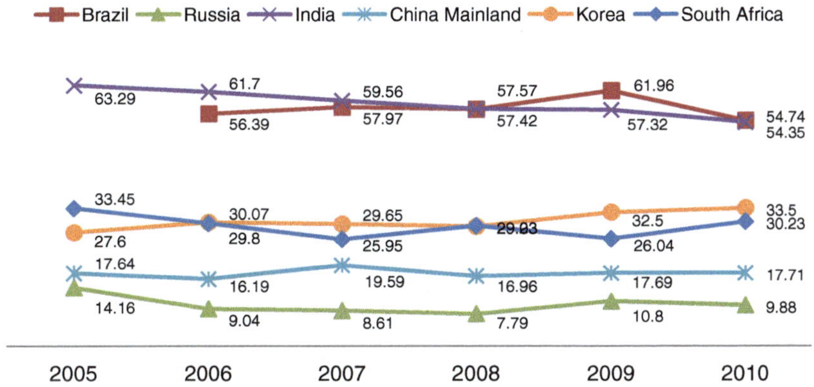

Fig. 2.2 Total general government debt (% GDP) of Brazil, Russia, India, China, Korea, and South Africa from 2005 to 2010

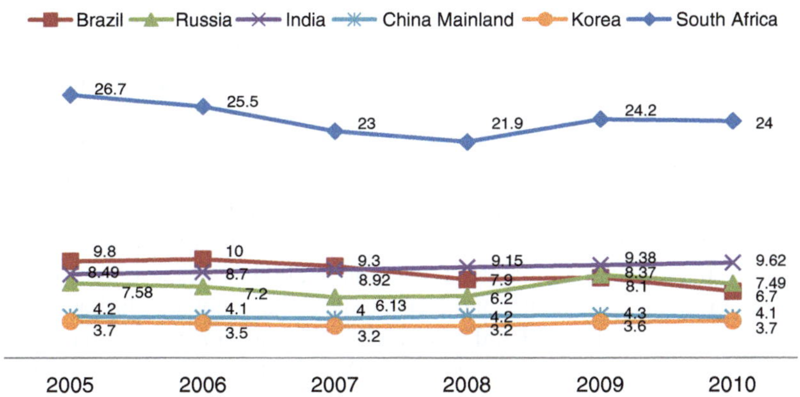

Fig. 2.3 Unemployment rate of Brazil, Russia, India, China, Korea, and South Africa from 2005 to 2010

9.88% in 2010, whereas China remains at the same debt level over the six years. In contrast to the obvious decline of real GDP growth in Fig. 2.1, the government debt levels of these six countries were not affected by this financial crisis, even with special stimulus packages.

Figure 2.3 indicates that the unemployment rates of these six countries are relatively stable, with the exception of South Africa in 2008. The figure shows three clusters as well. Over the years, China and Korea had unemployment rates of less than 5%; Brazil, India, and Russia had rates between 5 and 10%; and South Africa had a high rate of around 25%. Comparing the unemployment rate between 2005 and

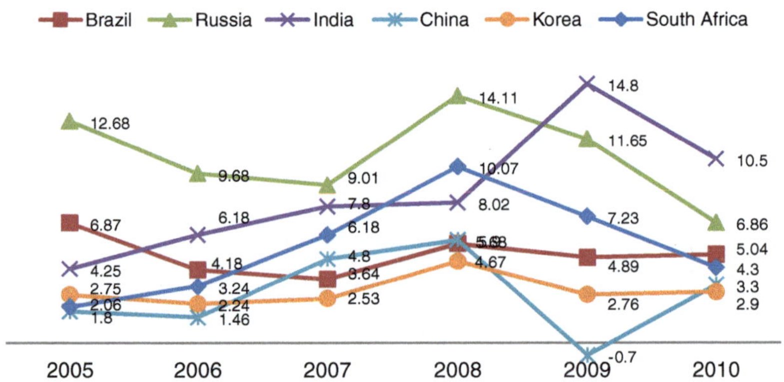

Fig. 2.4 Consumer price inflation of Brazil, Russia, India, China, Korea, and South Africa from 2005 to 2010

2010, Brazil and Russia experienced a reduction; China and Korea remained about the same; and India and South Africa had unemployment increases after the financial crisis. These statistics indicate that the employment of these countries, except India and South Africa, was not negatively impacted by this global financial crisis.

Figure 2.4 shows more ups and downs of consumer price inflation (CPI) among the six countries, compared to the previous three figures. The development pattern of CPI is similar for Brazil and Korea. CPI development for China and South Africa followed a similar path from 2005 to 2008, with South Africa having a higher rate than China. In particular China had a deflation in 2009, which was very likely due to excess production capacity and depressed global demand during the crisis (Branigan 2009). The deflation did raise slight concerns within the Chinese government before things returned to normal in 2010. In 2008, South Africa had recorded its highest CPI in six years. However, this was not sustained as it proceeded to experience two years of continuous decline in 2009 and 2010. As for India, it had the highest post-crisis CPI in 2009 and 2010 among the six countries. Russia's CPI had an early reaction to the financial crisis and rose to 14.11% in 2008, yet in 2010 it slid to 6.86% lower than its 2005 level. Examining the six countries together, India's CPI in 2010 was the highest of all and more than double its 2005 level.

In general, the above four figures for the BRICKS countries indicate that total general government debt and unemployment rates are relatively stable over the six years, whereas real GDP growth and CPI are more visibly impacted by the 2008 financial crisis with clear ups and downs. Among these six countries, India has the highest unemployment rate and consumer price inflation in 2010 after it recovered from the financial crisis. The fact that these two numbers have not resumed to their 2005 level for India while the other countries already have deserves some attention by Indian policy makers.

In what follows, we briefly describe the impact of the 2008 global financial crisis on the BRICKS countries. The depth of the report depends on the English literature available for each economy. For readers to gain a general picture about the efforts that each economy has expended to mitigate the negative impact of the financial crisis, we have summarized the details of stimulus packages implemented by the six countries in Appendix 1. Please note that the reported package is based on publicly available data and is not an exhaustive list. In addition, the reported amounts of stimulus packages were based on the exchange rate at the time of each stimulus, and thus vary. Readers can also refer to Appendix 2 for the important meetings conducted by key global leaders during the financial crisis.

Brazil

Brazil is the largest, most populous and the leading economic power in South America. After record growth in 2007 and early 2008, the onset of the global financial crisis hit Brazil in September 2008, as global demand for Brazil's commodity-based exports dwindled and external credit dried up (CIA 2012). However, Brazil only experienced approximately two quarters of economic recession and was one of the first emerging markets to begin recovering with improvement showing in the second quarter of 2009 (Mendonça 2010).

Before the crisis, Brazil had foreign reserves of US$205 billion in 2008 and property loans only accounted for 2.3% of the GDP in December 2008 (Cárdenas 2008; Mendonça 2010). Its banks had low foreign liabilities with only 30% of bank assets foreign-owned. In addition, all the banks have complied with an 11% capital requirement since 1995 (higher than the 8% ratio usually recommended by the Basel agreements). However, nearly half of Brazil's exports are commodities, which are sensitive to global supply and demand. As a result, based on the Economic Commission for Latin America and the Caribbean (ECLAC 2009) of the United Nations, between December 2008 and January 2009, the number of jobs fell by 755,000; between January and September of 2009, goods exports slid by 25.9% from the previous year; and exports of manufactured and semi-manufactured goods slumped by 32% and 30.8%, respectively. In addition, stock prices declined 20%, while spreads on Brazilian debt rose by more than 170 basis points (Cárdenas 2008).

In response to the downturn, Brazil's economic stimulus package focused on bringing inflation under control, generating higher investments, creating a floating exchange rate, and implementing prudent management of public finances and the fiscal measures for sustained economic growth. Its strong institutional framework and cautious monetary policy led to reduced inflation (Mendonça 2010). To boost investment, the government increased capital spending and investments in infrastructure and energy. It also started up an extensive program of government incentives and subsidies for housing construction (Cárdenas 2008). In an effort to limit the appreciation of the Brazilian currency, the government increased dollar reserves and capital controls.

There are three major factors that allowed Brazil to better resist the global downturn than many other countries. First, Brazil's recent sound policies and improved balance of payments made its economy more resilient to external impact (Gurria 2009). Second, the policies that centered on inflation, a floating exchange rate and prudent management during the crisis were appropriate (Gurria 2009). Third, Brazilian exports began to enjoy a positive benefits from the strong Chinese recovery. In 2009, China was the largest trade partner of Brazil and was the destination for 13% of its exports (Mendonça 2010).

In summary, the swift and comprehensive coping measures, including monetary and financial policy, exchange-rate and foreign-trade policy, sectoral policies, labor and social policies, and fiscal policy (Valadão and Porto 2009) of the Brazilian government helped minimize the negative impact of the financial crisis. Readers can refer to Appendix 1 for the details of Brazilian stimulus packages.

Russia

Russia has undergone significant changes since the collapse of the Soviet Union. It has moved from a globally isolated, centrally planned economy to a more market-based and globally integrated economy (CIA 2011). Russia enjoyed a decade of high, uninterrupted economic growth between 1999 and 2008, with annual real GDP growth averaging 6.9% during this period (Thessaloniki 2011). In 2009, Russia was the world's largest exporter of natural gas, the second largest exporter of oil, and the third largest exporter of steel and primary aluminum. However, this heavy reliance on commodity exports makes Russia vulnerable to the fluctuation of global commodity prices. As a result, its economy was one of the hardest hits by the 2008 global economic crisis as oil prices plummeted and the foreign credits that Russian banks and firms relied on dried up.

In addition, the Russian Central Bank owned around US$100 billion in mortgage-backed securities from the two American mortgage giants Fannie Mae and Freddie Mac that were taken over by the U.S. government afterwards (Delasantellis 2008). Consequently, Russian stock exchanges including MICEX and RTS lost 70% of their value and trading was suspended many times during the market turmoil. Russia's GDP contracted nearly 8% in 2009, industrial production tumbled by nearly 11%, plummeting exports collapsed by 36% and State revenues fell by almost 5% of GDP from 2007 to 2009 (Aslund et al. 2010). Losses were especially severe in finance and energy. To combat the economic downturn, Russia's initial fiscal policy response focused on supporting the financial sector and providing fiscal support for enterprises to revive export dependent industries. For instance, on September 17, 2008, the government provided the country's three biggest banks with US$44 billion (1.13 trillion rubles) loan in order to boost liquidity (Nicholson 2008). As the crisis from the financial sector spilled and started to afflict domestic demand, growth, and employment, the policy focus has begun shifting to households, infrastructure, and small- and medium-sized enterprises (Bogetic 2009).

What made things worse was that Russia began a military campaign against Georgia in response to Georgian ground attacks on South Ossetia in August 2008. This move was widely interpreted as Russian aggression and risky behavior, which resulted in capital outflows of around US$40 billion, thereby causing industrial production to decline nearly 14% year over year in March 2009 (Palmieri 2012).

To counter the crisis, the government enacted a stimulus package for fiscal measures implemented in 2008 amounting to more than US$113 billion (2.9 trillion rubles) or about 6.7% of GDP, larger than that implemented in most G20 countries (Bogetic 2009), thus increasing government expenditures from 33.7% of GDP to 40.6% of GDP (Aslund et al. 2010).

In total, the Central Bank of Russia spent one-third of its US$600 billion international reserves in late 2008 to slow the devaluation of the ruble (CIA 2011). Russia's stimulus policies can be largely categorized as macro coping policies and sectoral financial support. Specifically, macro-level policies included significantly increasing social welfare spending, cutting corporate tax profits by 4%, increasing tariffs on many imports, cutting the reserve requirement on liabilities, broadening the definition of eligible securities available for collateral, and expanding Russia's oil and gas assets in Europe and Asia for future growth and recovery. Sectoral financial support included buying out the foreign loan obligations of select companies, injecting billions of rubles into state controlled banks, and compensating certain banks for loan losses (Palmieri 2012).

Eventually, Russia's economy emerged from recession in the third quarter of 2009 after two quarters of negative growth (Nicholson 2009). According to the World Bank, Russia's strong short-term macroeconomic fundamentals made it better prepared than many emerging economies to deal with the financial crisis. Russian's recovery was also led by increased oil prices (which have risen from US$34/bl in early 2009 to around US$85/bl in 2010), stronger Russian domestic demand, a more flexible exchange rate to cope with the backdrop of oil price volatility, the moderation of rising unemployment by an increase in part-time employment and involuntary vacations as a temporary adjustment (Aslund et al. 2010).

As an energy rich country, Russia's economic rise and fall has become a global concern. In a BRIC-countries-focused knowledge forum conducted by the Wharton School at the University of Pennsylvania, Khema commented that the Russian government acted in a very organized, focused, and coordinated manner. This restored confidence in the banking system to some extent by announcing increased liquidity into the system, guaranteeing deposits, talking about cuts in oil export taxes and announcing significant investments in public infrastructure (Khemka et al. 2008). The forum concluded that the Russian government has been very proactive and done an excellent job projecting a clear plan and implementing that plan from the early stage of the crisis. In retrospect, another forum sponsored by Carnegie Foundation also reported that the Russian government's fiscal stimulus was an appropriate response to the shocks of the financial crisis, and resulted in the relatively quick recovery of the Russian economy (Aslund et al. 2010).

India

At the outbreak of the US sub-prime mortgage problems, much of Indian government's reaction to the crisis was based on the notion that it was a financial sector problem and that India's real sector was insulated and decoupled (Debroy 2010). The following facts further strengthen such a conception: First, Indian banks have strong balance sheets and are well capitalized and well regulated (Jha 2008). Second, India has a healthy external balance, with high levels of foreign exchange reserves (US$283.94 billion in October 2008) and low ratio of short-term external debt to GDP. Third, India's growth in the last few years has been driven predominantly by domestic consumption and domestic investment with external demand accounts of less than 15% of its GDP (Taindian News 2008; United Nations 2009).

However, the contagion still spreads to India because of its rapid globalization since the 1990s (Kumar and Soumya 2010a, b). India's two-way trade (merchandise exports plus imports) as a proportion of GDP grew from 21.2% in 1997–1998 to 40.6% in 2008–2009. As a result, India was still hard hit by the crisis, with its stock market index plummeting from 20,873 in early January 2008 to 9,093 in late November 2008, a 56% fall over a period of 11 months (Joseph 2009a, b). Equity markets were down by more than 50%, and both export sectors and real estate sector have taken a very severe downturn because of credit issues in the market (Khemka et al. 2008). Around one million jobs were lost after September 2008, especially in sectors like gems and jewelry, garments, leather, and handicrafts (Debroy 2010).

The domino effect started when foreign institutional investors withdrew funds from all over the emerging markets to meet the liquidity requirements of their principals in the U.S. As a result, in India, there was a substantial decline in net capital inflows in the first half of 2008–2009 to US$19 billion. This is significantly lower when compared to the US$51.4 billion India posted the same period a year before (Bajpai 2010a, b). The U.S., European Union and the Middle East, which together account for three quarters of India's goods and services trade, were in a synchronized downturn (United Nations 2009). Then, three major channels—the financial channel, the real channel, and the confidence channel (United Nations 2009)—worked through their ways to the business sectors and the general public and then crumbled the economy. When activities came to a halt all of a sudden, it had a tremendous effect on the Indian Stock Market and exchange rates due to the supply and demand imbalance within the foreign exchange market (Bajpai 2010a, b). Consequently, India's industrial growth faltered, inflation was at double-digit levels, current account deficit was widening, foreign exchange reserves were depleting and the rupee was depreciating (Chandrasekhar and Ghosh 2008).

In response to such unexpected turmoil, the first priority of India's Ministry of Finance was to reassure the people of the country's stability in the financial system and of the safety of bank deposits in particular. The thrust of the various policy initiatives by the Reserve Bank of India (central bank) focused on providing ample rupee liquidity and maintaining a market environment conducive for the continued flow of credit to productive sectors (United Nations 2009).

To further counter the negative fallout of the global slowdown on the Indian economy, the central government announced three successive fiscal stimulus packages. The first package came about in early December 2008, the second one in early 2009, and the third one in early March 2009. Readers can refer to Appendix 1 for the details. The packages include: across-the-board central excise duty reduction by 4 percentage points; additional spending of US$4.2 billion (Rs. 200 billion); additional borrowing by state governments of US$6.3 billion (Rs. 300 billion) for plan expenditure; and assistance to certain export industries in the form of interest subsidy on export finance and refund of excise duties/central sales tax (Joseph 2009a, b). In addition, the sharp rise in government consumption growth cushioned the sharp drop in aggregate demand and prevented a much sharper fall in GDP growth in the second half of 2009. In total, fiscal stimulus administered by India was around 6% of GDP (Joseph 2009a, b).

In addition to the government's proactive fiscal policy, Indian banks and financial institutions were not tempted to buy the mortgage-supported securities and credit-default swaps (Tharoor 2010), which also safeguarded the economy from being slashed by the toxic financial instruments. In spite of the economic downturn, India is still the second fastest growth country during the financial crisis, next to China only. In summary, financial stability in India was achieved through perseverance of prudent policies which prevented institutions from excessive risk taking, and financial markets from becoming extremely volatile and turbulent. As a result, financial markets remained orderly and financial institutions, especially banks, remained financially sound (United Nations 2009) throughout this crisis.

China

China is the fourth largest country in the world geographically after Russia, Canada, and the U.S. Currently, China is also the world's second largest economy after the U.S. after adjusting for differences in cost of living (CIA 2011). In the roughly five-year period before the start of financial crisis, China underwent drastic economic changes that allowed it to be an economic superpower in 2008. The initial step came on December 11, 2001, when China became a member of the World Trade Organization (WTO)—a milestone in its integration with the world economy (WTO 2012). Since then, underpinned by rapid expansions of trade and deep structural changes, its economy has continued to deliver yearly double-digit growth. The speed of the integration of China into the world economy, coupled with excess demand worldwide, pushed up China's current account surplus to as much as 11% of GDP by 2007.

After the financial crisis hit, thousands of private companies in China closed down. In the first half of 2009, exports sank 21.8% and imports declined 25.4% (Heilmann and Schmidt 2010). The Chinese government estimated that as of 2007, there were 286,200 approved foreign-invested companies in China. Such firms employed more than 42 million people and accounted for 31.5% of gross industrial output value (Morrison 2011). With the international money market crunch, capital flight from emerging countries is a common practice. The speed and scope of this hit came as a shock for Chinese policymakers. In an effort to mitigate the crisis, Chinese president Hu Jintao pledged an economic stimulus measure worth US$586 billion (RMB 4 trillion) in early November of 2008, which amounted to 15.5% of the country's GDP in 2007 (Heilmann and Schmidt 2010). In addition, China also implemented measures such as increasing investments in public infrastructure, loosening monetary policies to increase bank lending and providing various incentives to boost domestic consumption (Morrison 2011). It is important to note that the size of China's stimulus package is comparable to that of the United States', while its GDP is only a third as large (Fleet 2010). Herd et al. (2011a, b) also reported that China responded to the 2008 crisis with fiscal stimulus far greater than that of other OECD countries.

China launched its Economic Stimulus Plan mainly to expand the public sector, including pumping more public investments into infrastructure for rail network, roads, and ports development. It also created measures to increase affordable housing, lower taxes on real estate sales and commodities, and ease credit restrictions for mortgage as well as small and medium enterprises (Fleet 2010). China's stimulus funding differs significantly from that in other countries, the central government contributes only about one-fourth of all funds by issuing bonds over a two-year period from 2008 to 2010. The remaining three quarters were to be provided by local governments, state-controlled enterprises, and the market (i.e., government-linked financial institutions and nonpublic firms under government guidance) (Heilmann and Schmidt 2010). That is, a large part of the stimulus package involved off-budget expenditure by local authorities, which resulted in the public expenditure rising by nearly 3% of GDP in 2009.

Critics pointed out that the share of the package was aimed at major infrastructure projects, as opposed to direct stimulus of consumption and the RMB 4 trillion stimulus package was more a policy than a package (Fleet 2010). Nonetheless, the package worked and recovery started to show by the second half of 2009. This timeframe preceded that of other major economies. As a result, companies were hiring again, restaurants were filling up and consumption statistics were turning upward. After the crisis, both the estimated output gap and the OECD composite leading indicators suggested that China is operating above capacity with its ongoing rapid economic transformation (Herd et al. 2011a, b).

Looking back at the stimulus decision-making process, the experience of focusing on public expenditure in the 1997 Asian Financial Crisis gave the 2008 Beijing policymakers valuable lessons to draw from. Beijing designed more than 50% of the entire RMB 4 trillion in the initial stimulus package to focus on infrastructure. In this manner, the money could be spent quickly and had a significant impact on employment where it was needed the most since many low-skilled workers had lost low-end export-processing jobs. In reality, much of the money actually found its way to private firms who were subcontractors to the initial state-owned enterprise beneficiaries, and it had a huge impact on consumer confidence (Fleet 2010). In addition, China included in its stimulus package a massive increase in liquidity; that is, funds from the national government would not exceed around a quarter of the designated RMB 4 trillion. Also, the package stipulated that 40% of the amount would be provided by bank lending matched with reduced reserve requirements and interest rates cut (Heilmann and Schmidt 2010). Furthermore, the measures consisted of substantial regulatory change, including a lowering of down payments required for mortgages, reduction or elimination of value added tax (VAT), lower fuel prices and subsidies for smaller cars (fuel efficiency), and an expansion of subsidies for consumer goods purchases in rural areas.

In a knowledge forum conducted by the Wharton school at the University of Pennsylvania, Chou (Khemka et al. 2008) commented that the financial crisis was not a bad thing for China as its economy was growing too fast and needed to be held in-check. The preoccupation of the Chinese government for the last few years before the crisis had been how to cool down the economy and the real estate prices, which had become out of the reach of the common people. As far as China can, it had been trying to move away from being overly reliant on the exports sector of the economy. Therefore, the financial crisis offered an opportunity for China to adjust its economy to be more independent and resilient to future external impact.

In general, the fast and decisive government stimulus measures worked well in China. In addition to its own recovery, China also helped pull the world economy along during this financial crisis (Fleet 2010).

Korea

The Republic of Korea has been one of the most successful emerging economies in recent decades (Pascha 2010). In Korea, private investments have typically supported the export economy since the 1960s, which resulted in exports of goods accounted for about 45% of GDP in 2008 and commercial services added an additional eight percentage points. Yet, even before the 2008 financial crisis, Korea was experiencing some macroeconomic problems, such as the high value of imports driven by higher prices for raw materials and energy. Job growth, domestic consumption, investment and corporate profits were all negatively affected by such development before the financial crisis set in.

At the onset of the crisis, Korea felt rather safe, because of its significant accumulation of foreign exchange reserves (US$240 billion by September 2009), and the relatively sound financial situation of Korean banks and enterprises as compared to 1997 during the Asian Financial Crisis. However, due to the tightening of international liquidity, the government had to step in and promise a US$30 billion infusion to support the banking system (Pascha 2010).

The global financial crisis hit the Korean economy through the sudden reversal of capital flow, which dried up the domestic and international liquidity. In the first six months of 2008, net FDI in South Korea turned negative for the first time since 1980 (OECD 2008). Capital outflow, which amounted to US$54.5 billion in 2008, reduced the supply of foreign currency to the point that the exchange rate of the

Korean currency (won) lost over 50% of its value from the beginning of 2008 to the fourth quarter of 2009. In addition, the global contraction of demand reduced Korea's export by over 40% in the fourth quarter of 2008 (Yoon 2011).

Although the exposure of the Korean financial system to toxic financial assets was small, banks and enterprises were seriously affected by the international liquidity crunch and by the steep drop in export markets. An important aspect of Korea's crisis management was the stabilization of foreign exchange and currency markets. To this end, the Bank of Korea extended its existing swap agreements with the U.S., China, and Japan beyond their normal limits, amounting to US$90 billion (Kim 2010c; Yoon 2011). In October 2008, the government responded by pledging more than US$100 billion in loan guarantees and an infusion of US$30 billion to prop up the Korean banking system, hoping that the additional liquidity would help Korean banks repay or roll over the banks' US$80 billion in foreign currency loans that came due by June 2009 (OECD 2008). In addition, the government provided US$55 billion of liquidity directly into the foreign exchange market as a credit mostly for trading companies to preserve their capacity to earn foreign currency (Yoon 2011). In the case of fiscal policies, the government expanded fiscal budget by more than US$355 billion (50 trillion won, over 5% of GDP) since the second half of 2008, which included tax refund on oil consumption, fiscal spending based on supplementary budget in 2008, tax cuts, additional fiscal spending, and supplementary budget in 2009 (Kim 2010c).

Internally, for 2008 and 2009, Korea's total fiscal stimulus package amounted to US$352 billion (49.6 trillion won), including US$236 billion (33.2 trillion won) of public expenditure and US$116 billion (16.4 trillion won) in tax reduction. The volume accounted for 4.5% of the GDP in 2009 (Yoon 2011). The largest stimulus package launched by the Korean government to boost public infrastructure was the Green New Deal. This comprised nine major projects totaling US$36 billion (spread over four years), including around US$6 billion to improve energy conservation in villages and schools, US$7 billion on mass transit and railroads, and almost US$11 billion on river restoration (Pascha 2010). In addition, special tax deductions were given to solar cell manufacturing plants, the cleaning of Korea's four biggest rivers and the erection of flood defenses. The Green New Deal met the social expectation that the stimulus scheme should create jobs. According to the plan, 960,000 positions were to be created within four years, with 140,000 of them coming in 2009 (Pascha 2010). Since most of these jobs are for manual labor, the Green New Deal was also considered as social policy on behalf of the weak and potentially underemployed. In parallel to the announcement of this project in January 2009, the Korean government announced its intention to support the so-called "new growth engines" of 17 specific industries including sustainable energy, information technologies, health care, and tourism (Pascha 2010).

Support for specific industries included helping the automobile industry in the form of a tax deduction of 30% (offered for a limited period) and setting up the Bank Recapitalization Fund in February 2009, in addition to the Bond Market Stabilization Fund established in December 2008. As part of a tax deduction for investments, a more favorable rate was also offered for investments in provincial areas, which indirectly represented support for SMEs. As for consumer spending, the most important measures were the personal income tax reduction and the support for low-income households. Business benefits most from corporate tax reduction and from measures aimed at increasing the capital supply in the banking system. Readers can refer to Appendix 1 for the details.

Even though Korea was one of the G-20 economies hit hardest by the financial crisis (Pascha 2010), it was the first OECD country to escape from negative economic growth. By the summer of 2009, the IMF had revised Korea's growth projection for 2009 from −3% to −1.75% (Pascha 2010). Primary reasons include: first, Korea had better initial conditions than other economies because of reform measures after the 1997 Asian Financial Crisis; second, Korea had an international network to establish swap arrangements of US$90 billion to stabilize its foreign exchange market (Yoon 2011); and third, the Korean government took effective and bold measures in a prompt manner. In addition, the increasing demand from its major trade partner—China was another major factor in Korea's recovery since the second quarter of 2009 (Kim 2010a, b). According to OECD (2010), the magnitude of fiscal stimulus package implemented by the Korean government in responding to the global financial crisis is 6.1% of GDP, the highest among OECD member countries adopting explicit crisis-driven stimulus programs (Kim 2010a, b).

South Africa

South Africa is a middle-income, emerging market with an abundant supply of natural resources. It also sports well-developed financial, legal, communications, and transport sectors with modern infrastructures supporting relatively efficient distribution of goods to major urban centers throughout the region (CIA 2012). Since its turn towards democracy in 1994, South Africa has improved its reputation as the leading economic power in Africa and has increasingly attracted global investors (Draper et al. 2009). Continuous integration into the global economy has led to notable increases in its productivity (Heritage 2012). South African growth was robust from 2004 to 2007 as the country reaped the benefits of macroeconomic stability and a global commodities boom (CIA 2012; OECD 2010).

However, its economy began to slow in the second half of 2007 due to a serious electricity crisis (CIA 2012). The 2008 global downturn struck South Africa after it had already passed its boom, with the economy slowing sharply as the country was experiencing its first recession in 17 years (OECD 2010). Thus, South Africa entered the crisis with a greater degree of vulnerability, having a large current account deficit, high interest rates, and high inflation (Padayachee 2012). The high interest rate halted growth in private consumption (OECD 2010).

South Africa's decline was led by manufacturing and mining where the sudden drop in export demand was reflected in a sharp reduction in private investments and subsequently in falling employment. Its GDP figure for the last quarter of 2008 came in negative, with manufacturing falling by 21.8%, the automobile industry (the main contributor to international trade tax revenues) was down over 30% year-on-year, and mining production continued to fall as global commodity prices remained depressed (Zini 2009). As a result, job creation was slowing down, consumers' expenditure declining, credit extension to the private sector shrinking, and housing prices dropping (Zini 2009). Unemployment was already high prior to the crisis and began rising further starting from the fourth quarter of 2008 (ILO 2010). Consequently, the country experienced a sharp deceleration of growth, dropping from more than 5% in 2007 to almost 2% in 2009 (OECD 2010). The Johannesburg Securities Exchange All-Share Index fell from a high of 32,542 on May 23, 2008 to a low of 18,066 on November 21, 2008 (Padayachee 2012). From October 2008, the nation suffered from capital withdrawal and a depreciation of its currency, the Rand (Draper et al. 2009). In December 2008 tripartite negotiations began among organized labor, business, and government to formulate a collective response to the crisis (ILO 2010).

Nevertheless, South Africa's decline in output was moderated by a countercyclical policy response, made possible by past fiscal prudence and by the resilience of the non-impacted banking system (OECD 2010). Furthermore, to remedy the problems, the government's policy response was prompt and focused on housing, education, social protection, public works, and health programs under the expansionary budget (Zini 2009). In addition, the Training Layoff Scheme, rolled out in September 2009, provides skills training and an allowance to employees during a negotiated layoff period, although they technically remain employed (ILO 2010).

According to OECD (2010), South Africa's downturn was fairly shallow: the peak-to-trough fall in output was smaller than in most OECD countries and emerging market economies. Its strong banking system supports the economy well, such as bankruptcy laws favoring creditors in recovering collateral for bad loans and conservative approaches on the part of both the regulator and the bank themselves (OECD 2010). Specifically, the strength of South Africa's banking sector includes strong profitability even during the financial crisis, low level of non-performing loans with comfortable capital cushions, and lack of direct exposure to problematic assets from the U.S. and Europe. With healthy foreign reserves and continuous financial inflows, South Africa has no problem to fund the deficit of around −3.9% in 2009–2010 (Zini 2009). Therefore, despite the South African economy having been hard hit by the global financial crisis, no exchange rate shock or "sudden stop" was experienced (Draper et al. 2009).

Another important factor that mitigated the impact of the crisis was the previously approved ambitious capital expenditure plan in energy and transport by public enterprises. This plan was not revised downward, which greatly supported output during the decline. Monetary policy was also eased, but not particularly aggressively. During the financial crisis, no emergency actions, such as capital support for banks or quantitative easing to support lower interest rates, were needed (OECD 2010).

Chapter 3
National Intellectual Capital Development of the Six BRICKS Countries

In this chapter, we first present the development of national intellectual capital in the six BRICKS countries over the 6 years and then graphically display the relationship between each individual capital with GDP per capita (ppp). Afterwards, we describe the relative position of long-term and short-term intellectual capital in these six countries and explain the dynamics of NIC for three time periods (2005–2006, 2007–2008, and 2009–2010). Finally, we illustrate the NIC trajectory of these six countries on three-dimensional (3D) 48-country landscapes.

National Intellectual Capital Development

The NIC measurement model in this study is based upon the indicators described in Appendix 3, containing human capital, market capital, process capital, renewal capital, and financial capital. This model has been statistically validated for its validity and reliability. The definitions of the indicators are listed in Appendix 4. This booklet series covers a total of 48 countries in 11 country clusters (see Appendix 5 for details).

For a full picture of all the 48 countries, Appendix 6 exhibits the scores and rankings of human capital, market capital, process capital, renewal capital, financial capital, and overall NIC covered in this booklet series. Table 3.1 summarizes the status of the six BRICKS countries compared with the total 48 countries reported in Appendix 6. Appendix 7 provides the individual economic profiles with additional statistics. For the NIC ranking (Table 3.1), Korea ranks 22, followed by South Africa at 37, Russia at 38, China at 40, Brazil at 42, and India at 44.

Among the five types of capital, Korea is comparatively better in human capital (#19) and renewal capital (#19). China and India are relatively better in market capital, ranking #28 and #20, respectively.

The following six figures compare human capital, market capital, process capital, renewal capital, financial capital, and NIC for the six countries. All individual capital scores are based on 1–10 scale with the NIC score being the sum of the five capital scores.

C.Y.-Y. Lin et al., *National Intellectual Capital and the Financial Crisis in Brazil, Russia, India, China, Korea, and South Africa*, SpringerBriefs in Economics 18,
DOI 10.1007/978-1-4614-6089-3_3, © The Author(s) 2013

Table 3.1 National intellectual capital ranking of Brazil, Russia, India, China, Korea, and South Africa (2005–2010)

	Human capital		Market capital		Process capital		Renewal capital		Financial capital		NIC	
Country	Score	Ranking	Score	Ranking	Score	Ranking	Score	Ranking	Score	Ranking	Score	Ranking
Brazil	4.757	44	4.589	38	3.076	46	1.753	36	8.481	40	22.656	42
Russia	5.530	33	3.860	46	3.588	41	2.349	28	8.822	31	24.150	38
India	3.926	48	5.620	20	3.556	42	1.954	34	7.285	48	22.340	44
China	4.942	39	5.184	28	3.746	40	2.099	33	7.967	44	23.938	40
Korea	6.964	19	5.123	29	5.282	26	4.595	19	9.383	25	31.348	22
South Africa	4.708	45	4.976	30	4.472	31	1.677	39	8.472	41	24.306	37

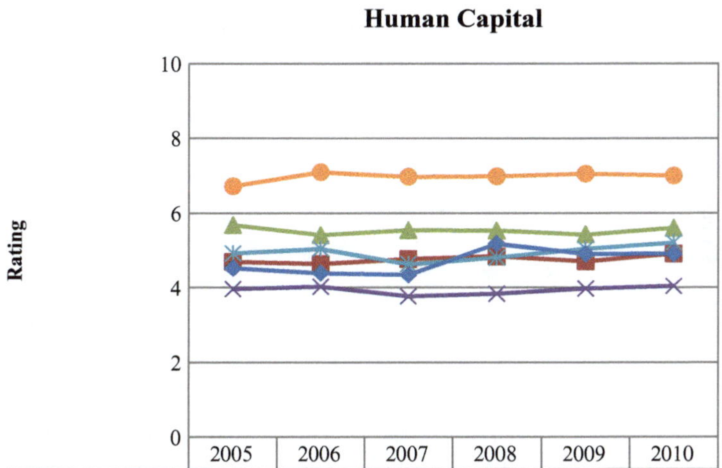

	2005	2006	2007	2008	2009	2010
Brazil	4.69	4.64	4.76	4.84	4.71	4.91
Russia	5.67	5.41	5.54	5.53	5.43	5.60
India	3.96	4.01	3.76	3.83	3.96	4.04
China	4.92	5.04	4.63	4.81	5.04	5.21
Korea	6.72	7.09	6.96	6.98	7.05	6.99
South Africa	4.52	4.39	4.34	5.18	4.91	4.91

Fig. 3.1 Human capital of Brazil, Russia, India, China, Korea, and South Africa

Human Capital

National human capital represents a nation's investment in nurturing qualified human resources, such as higher education enrollment, skilled labor, and public expenditure in education. Human capital is a long-term oriented NIC (Lin and Edvinsson 2011: 323), which takes time to develop and is not susceptible to external impact. Figure 3.1 shows that human capital of these six countries remained relatively stable over the 6 years. South Africa was the only exception, posting good progress in 2008.

In 2010, Korea has the highest human capital, followed by Russia, China, South Africa and Brazil together, and then India. Yet, human capital score of Korea is far above that of the other five countries. In 2010, India and Russia maintained very similar human capital score in line with their 2005 numbers. The other four countries took a step further and posted good progress during this 6-year period.

Market Capital

National market capital represents international trade and relationships such as cross-border ventures, globalization, and exports of goods. Figure 3.2 shows that in

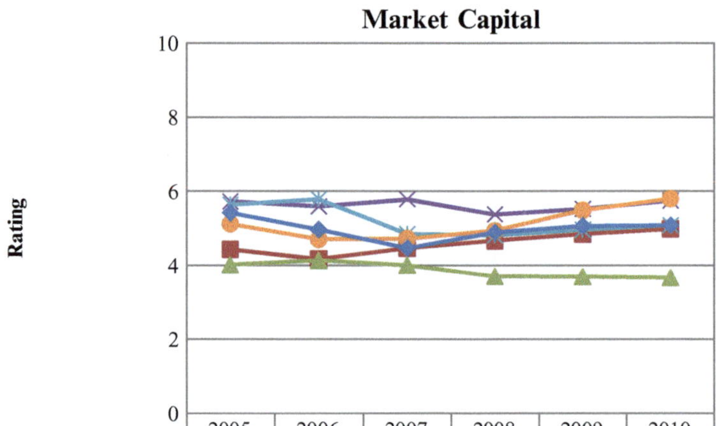

	2005	2006	2007	2008	2009	2010
Brazil	4.43	4.17	4.46	4.66	4.84	4.98
Russia	4.01	4.13	3.99	3.69	3.68	3.65
India	5.72	5.58	5.78	5.38	5.52	5.74
China	5.63	5.78	4.85	4.81	4.95	5.08
Korea	5.12	4.70	4.72	4.93	5.48	5.79
South Africa	5.41	4.96	4.46	4.88	5.06	5.08

Fig. 3.2 Market capital of Brazil, Russia, India, China, Korea, and South Africa

2005, there were two groups, Brazil and Russia, in the low market capital group, with the other four countries in the high market capital group. In 2010, the distribution split into three groups; Russia remained in the low group, Brazil, China and South Africa in the middle group, and India and Korea in the high group. It can be seen that Brazil and Korea increased their market capital over the 6-year period, India remained steady, and the other three countries decreased in market capital. From 2005 to 2010, Brazil had continuous market capital increases. On the other hand, Russia had the continuous decrease in the same time period. India had ups and downs and China showed an early decline in 2007. Korea had an even earlier decline in 2006, but then rebounded stronger in 2009 and 2010 to above 2005 levels. Market capital of South Africa drops from 2006 to 2007, but then rebounded little by little since 2008.

Process Capital

National process capital represents the infrastructure required for building a prosperous society, such as capital availability, mobile phone subscribers, and government efficiency. Figure 3.3 indicates that Korea has the highest process capital, followed by South Africa, India, Russia, China, and Brazil. China is the only country that had process capital decrease over the 6-year period. In 2007, all the countries

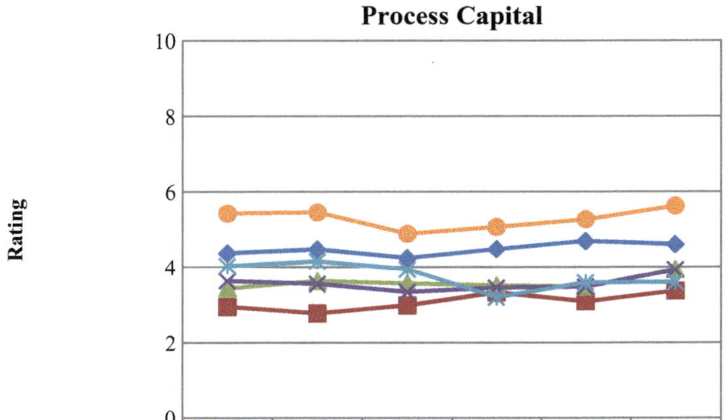

	2005	2006	2007	2008	2009	2010
Brazil	2.94	2.77	2.98	3.32	3.09	3.36
Russia	3.43	3.62	3.56	3.51	3.47	3.94
India	3.63	3.54	3.33	3.44	3.47	3.92
China	4.03	4.15	3.93	3.19	3.58	3.59
Korea	5.42	5.45	4.88	5.06	5.26	5.62
South Africa	4.37	4.47	4.24	4.48	4.68	4.60

Fig. 3.3 Process capital of Brazil, Russia, India, China, Korea, and South Africa

showed declining process capital with the exception of Brazil. In 2008, only Russia and China continued to have process capital decline whereas every other country had reversed this trend. Over the 6-year period, Korea was consistently the leader in process capital, followed by South Africa. On the other end of the spectrum, Brazil came up last in process capital.

Renewal Capital

National renewal capital represents the capability of R&D and innovation, such as R&D spending, scientific articles, and patents. Figure 3.4 shows that Korea is at the upper end of the scale and has kept on progressing over the years. The other five countries cluster together at the lower end. In 2006, South Africa had an early small scale renewal capital decline but gradually rebounded back to its 2005 level in 2010. In 2008, only China exhibited renewal capital decline, yet China was able to successfully rebound in 2009 and 2010. On the contrary, Brazil, Russia, and India showed late renewal capital declines in 2009. Among these three countries, Brazil and Russia rebounded in 2010, yet India's renewal capital kept on falling in 2010. Over the 6-year period, Russia and India had renewal capital decreases, South Africa remains almost the same, and the other three countries had increases.

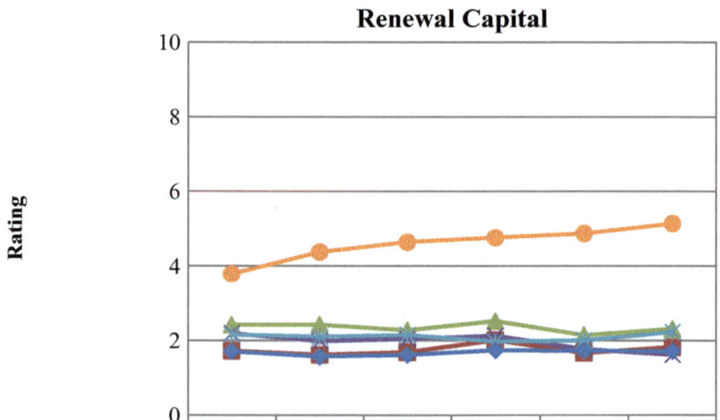

	2005	2006	2007	2008	2009	2010
Brazil	1.72	1.61	1.68	2.01	1.67	1.83
Russia	2.42	2.42	2.28	2.52	2.15	2.31
India	2.19	1.98	2.03	2.13	1.77	1.62
China	2.15	2.09	2.15	1.97	2.00	2.23
Korea	3.79	4.37	4.64	4.76	4.88	5.14
South Africa	1.70	1.57	1.61	1.73	1.73	1.72

Fig. 3.4 Renewal capital of Brazil, Russia, India, China, Korea, and South Africa

Financial Capital

Financial capital is representative of a country's GDP per capita (ppp) in US dollars transformed to a score on a scale of 1–10. The process of score transformation is explained in Appendix 3. Figure 3.5 shows that Korea was consistently the leading country, followed by Russia, Brazil, South Africa, China, and India. In actual dollars amounts, the average GDP per capita (ppp) of 2005–2010 was US$9,982 for Brazil, US$14,487 for Russia, US$2,745 for India, US$5,817 for China, US$26,589 for Korea, and US$ 9,871 for South Africa, respectively. In general, the financial capitals of these six countries were not impacted by the 2008 global financial capital, maintaining its stability throughout.

NIC

NIC is the overall national intellectual capital comprising the five types of capital described in the sections above. Figure 3.6 presents the overall trends of NIC. Again, Korea is the leading country in terms of NIC. It is way above the other five countries. China is the only country with NIC decrease over the 6-year period. India had three consecutive years of NIC decline (2006–2008) and South Africa had 2 years (2006–2007). Korea was the first to rebound in 2008.

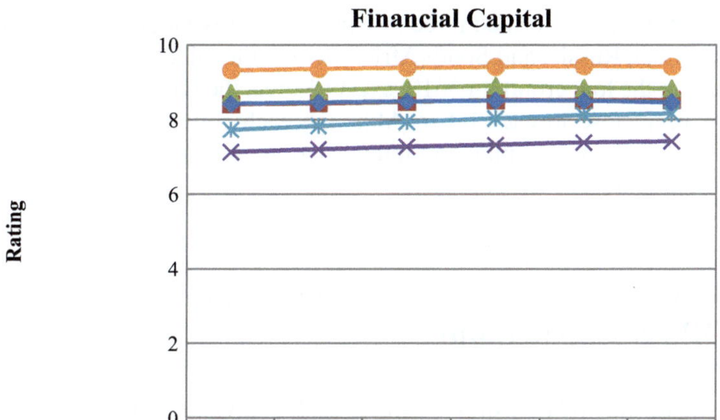

	2005	2006	2007	2008	2009	2010
Brazil	8.41	8.44	8.47	8.52	8.52	8.52
Russia	8.71	8.78	8.85	8.90	8.85	8.84
India	7.13	7.20	7.27	7.32	7.39	7.41
China	7.73	7.83	7.94	8.03	8.12	8.16
Korea	9.32	9.35	9.38	9.41	9.43	9.42
South Africa	8.42	8.45	8.49	8.51	8.50	8.46

Fig. 3.5 Financial capital of Brazil, Russia, India, China, Korea, and South Africa

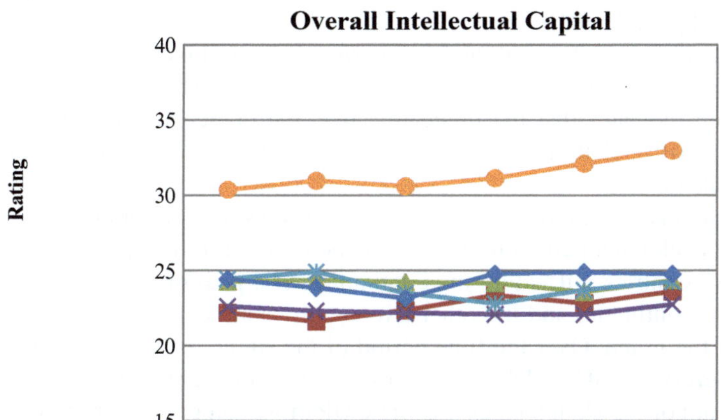

	2005	2006	2007	2008	2009	2010
Brazil	22.19	21.62	22.35	23.34	22.82	23.60
Russia	24.25	24.36	24.22	24.15	23.57	24.34
India	22.62	22.32	22.17	22.09	22.10	22.74
China	24.45	24.90	23.50	22.81	23.70	24.26
Korea	30.36	30.96	30.58	31.13	32.09	32.96
South Africa	24.42	23.84	23.14	24.78	24.88	24.77

Fig. 3.6 NIC of Brazil, Russia, India, China, Korea, and South Africa

In 2005, NIC scores formed three groups, Korea was far ahead of the other five countries; China, Russia, and South Africa were in the middle group; and India and Brazil were in the low group. In 2010, except Korea, the NIC of the BRICKS countries were very close to each other. Although each country has its own strength (Russia excels in human capital and India is strong in market capital), their overall NIC turns out to have small gap only.

Relationship Between Each Individual Capital and GDP Per Capita (ppp)

We investigate the relationship between each individual capital and economic growth by using the following six figures (Figs. 3.7, 3.8, 3.9, 3.10, 3.11, and 3.12). The graphs have four dimensions: the *X* axis changes from graph to graph (NIC, human capital, market capital, process capital, and renewal capital), the *Y* axis is a fixed dimension of GDP per capita (ppp) in US dollars (hereafter termed GDP), the color shade represents renewal capital with red denoting higher levels of renewal capital, and the bubble size denotes the population of an economy. Figure 3.7 shows the relationship between NIC and GDP for all 48 countries. Figures 3.8, 3.9, 3.10, 3.11, and 3.12 display the co-development of the intellectual capital-GDP relationship over the 6-year timeframe for the six countries. Since financial capital is the transformed 1–10 score from GDP, there is no need to plot its relationship against itself. In the figures, 2005 is the starting point tagged with the name of the country, and the end point is 2010.

In particular, Fig. 3.7 is plotted to clearly show a full picture of the relationship between NIC and GDP for 48 countries in year 2010. The strong positive correlation indicates that the higher the NIC the higher the GDP. Consequently, enhancing NIC is definitely a strategic issue that deserves the attention of national policy makers.

With the data of all 48 countries, Fig. 3.8 shows the formation of the NIC and GDP co-development continuum from the bottom left to the upper right. All paths of the six countries fall within the continuum. The path of Korea is at the middle part of the continuum and the other five countries are clustered at the left bottom corner of the graph. Over the time period of the study, both Brazil and Korea had good progress in NIC and GDP, although at different levels of development. The vertical path of Russia indicates that the GDP of Russia grew over the years, yet its NIC did not have much improvement.

Figure 3.9 shows the "human capital-GDP" continuum formation of the 48 countries and the co-development paths of the six countries. The paths of all countries, except India, are within the continuum. Again, Brazil and Korea have good co-development of human capital and GDP, whereas the human capital of Russia declined even though it had continuous GDP growth. The path of China and India shows some human capital-GDP progress, yet on a relatively small scale. South Africa had more human capital improvement than GDP growth over the 6 years.

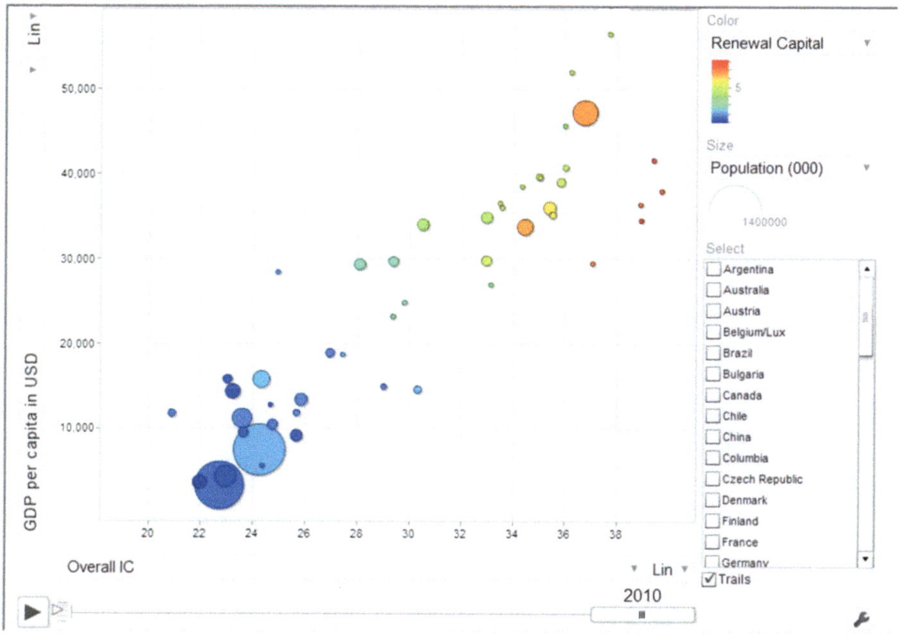

Fig. 3.7 NIC vs. GDP per capita (ppp) for 48 countries in 2010

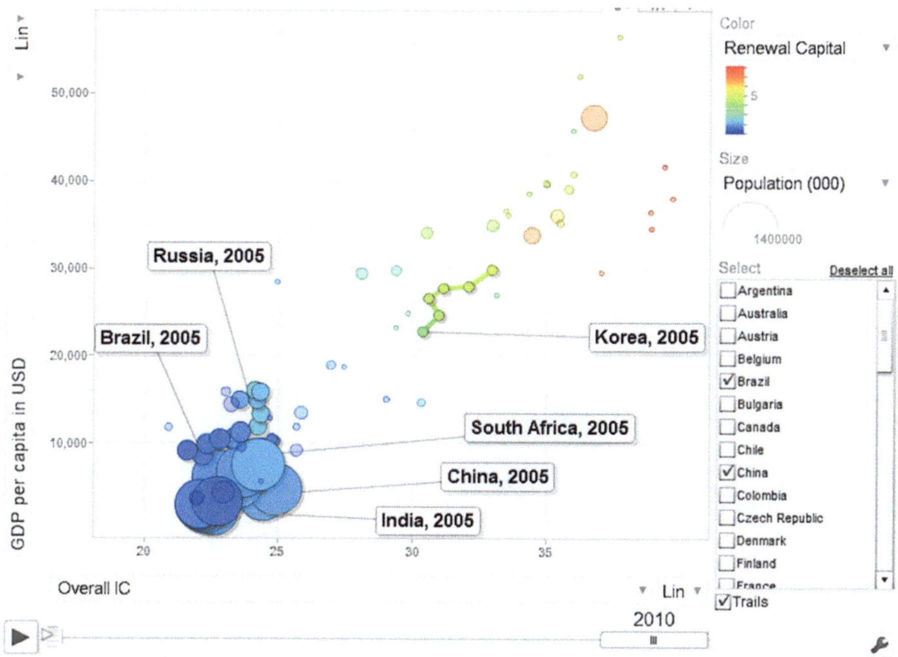

Fig. 3.8 The development of NIC and GDP per capita (ppp) for the BRICKS countries from 2005 to 2010

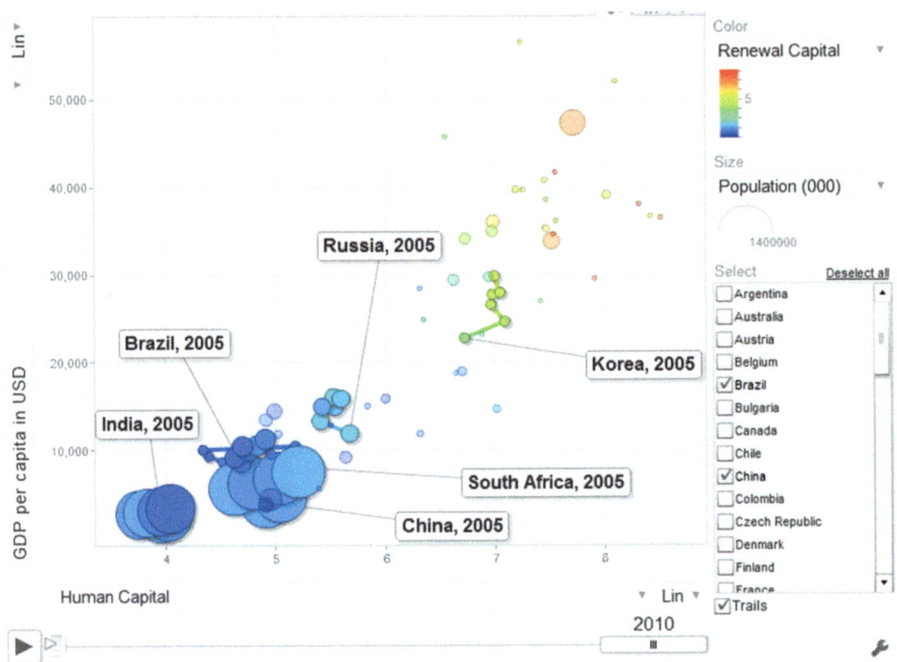

Fig. 3.9 The development of human capital and GDP per capita (ppp) for the BRICKS countries from 2005 to 2010

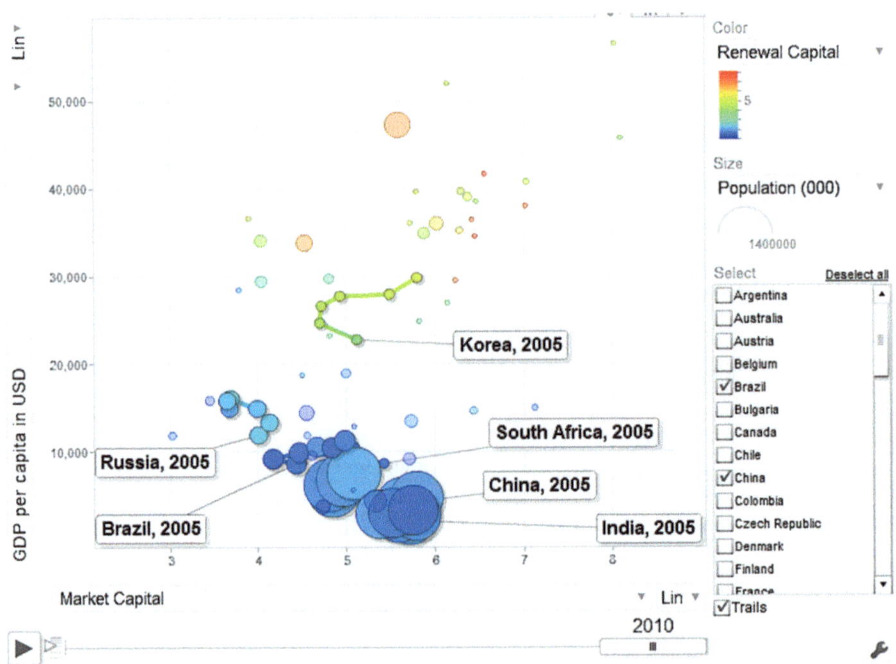

Fig. 3.10 The development of market capital and GDP per capita (ppp) for the BRICKS countries from 2005 to 2010

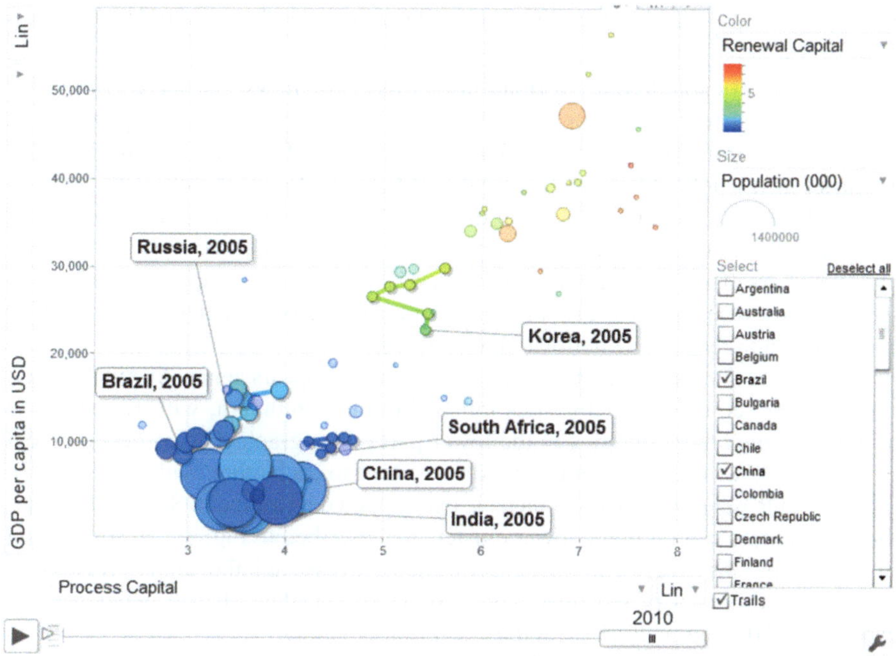

Fig. 3.11 The development of process capital and GDP per capita (ppp) for the BRICKS countries from 2005 to 2010

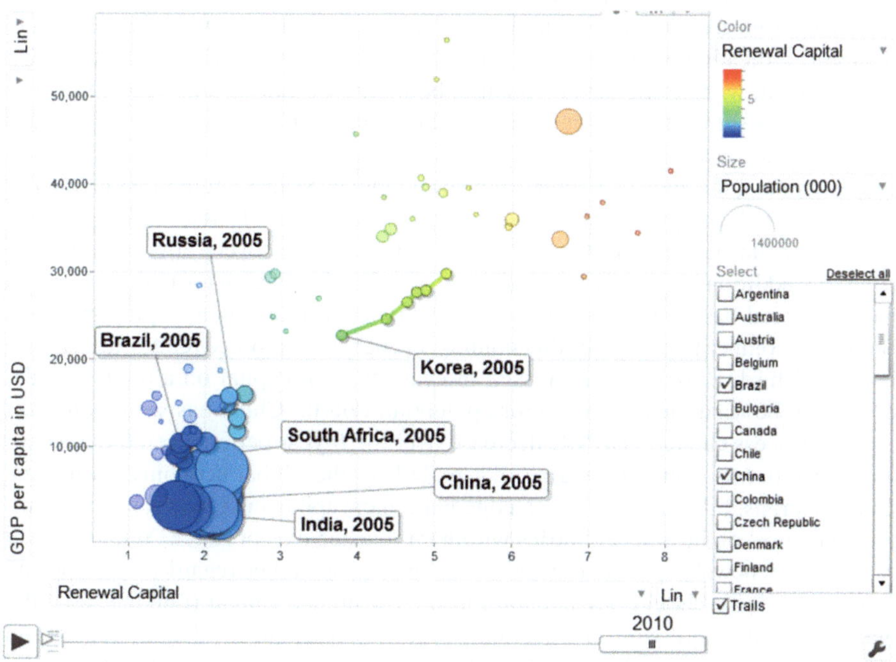

Fig. 3.12 The development of renewal capital and GDP per capita (ppp) for the BRICKS countries from 2005 to 2010

The "market capital-GDP" formation of Fig. 3.10 shows that there is no clear continuum like the last two figures as the data points spread further apart, which means that the correlation between market capital and GDP is not high for these 48 countries as a whole. The market capital-GDP co-development path of Korea is in the middle-middle, and the rest are in the middle-low part of the graph. Russia had a higher GDP than the other emerging countries, yet its market capital decreased over the years. This figure clearly shows that in 2005, market capital of China and India was even higher than that of Korea, indicating an area emerging economies can focus on in facilitating national development.

Figure 3.11 shows that the belt-shaped continuum appears again. All six "process capital-GDP" co-development paths fall within the continuum, with Korea in the middle and the other countries in the lower part of the graph. The co-development paths of the five emerging countries are very close to each other, with Brazil catching up with Russia. India and China have overlapped development. South Africa has the best process capital in the lower cluster. As described in Fig. 3.3, at some point of time during the 6 years, all the countries had process capital declines, very likely reflecting the impact of the financial crisis. However, they all had better process capital in 2010 than in 2005, with the exception of China.

Figure 3.12 shows that there is no clear continuum for the 48 countries. The "renewal capital-GDP" co-development path of the six countries is very different from the previous figures. They spread out in different parts of the graph. In general, more advanced countries gravitate towards the upper right corner. The less developed countries are crowded in the left bottom corner, tilted toward the GDP side (*Y* axis), indicating very low renewal capital. However, Korea has continuous advancement, with strong development in both renewal capital and GDP. The other five countries showed some renewal capital decline over the years, with India having the greatest decline.

In summary, the above six figures convey certain messages about the state of NIC and GDP co-development of various combinations. First, Korea is ahead of the other five countries in every type of capital with a good recovery after the financial crisis. In addition, it serves as a role model of national transformation. Second, human capital has more variation than other types of intellectual capital and may become a competitive edge in this country cluster, as it closely relates to national renewal. Third, Brazil has progressed in every type of capital on a modest scale, which can be accumulated into future potential. Fourth, China has seen its market capital, process capital, and NIC decrease over the 6-year period, even though economically it was a growth engine that pulled up the global economy during the financial crisis. Fifth, Russia has an edge in human capital and renewal capital comparing to most of the other countries within this cluster, except Korea. Yet, its market capital and renewal capital continued to decline and should be regarded as a warning sign. Sixth, India should pay attention to its low human capital (ranking last), flat market capital progression, and declining renewal capital. Its low human capital and renewal capital may be related to the high unemployment rate even after the financial crisis, as unemployment normally affects low-skilled labor.

Long-Term and Short-Term National Intellectual Capital

In addition to the six figures shown above that depict the relationship of each individual capital with GDP, Figs. 3.13 to 3.16 further plot the relationships between human capital and renewal capital, as well as between market capital and process capital. In our first study (Lin and Edvinsson 2011: 324), we found that the combined increase in human capital and renewal capital can lead to a nation's long-term development; whereas the combined increase in market capital and process capital might result in more immediate economic development. To present their relationships, Figs. 3.13 to 3.16 map the four capitals on a coordinate with two types of presentation. Based on the same data, Figs. 3.13 and 3.15 exhibit a scatterplot for the six countries only, whereas Figs. 3.14 and 3.16 show all 48 countries in the background so that readers will know the relative positions of these six countries. Please note that the connotation of color shade for Figs. 3.14 and 3.16 has been changed from renewal capital to GDP per capita (ppp).

Both Fig. 3.13 and 3.14 indicate that Korea is ahead of the other five countries, followed by Russia, China, Brazil/South Africa, and then India. Figure 3.13 particularly shows that Korea and China are advancing over the 6-year period, Russia and India are going backwards, and Brazil and South Africa have ups and downs in these two long-term oriented human capital and renewal capital. The vertical progression (except Korea) of Fig. 3.14 indicates good improvement in human

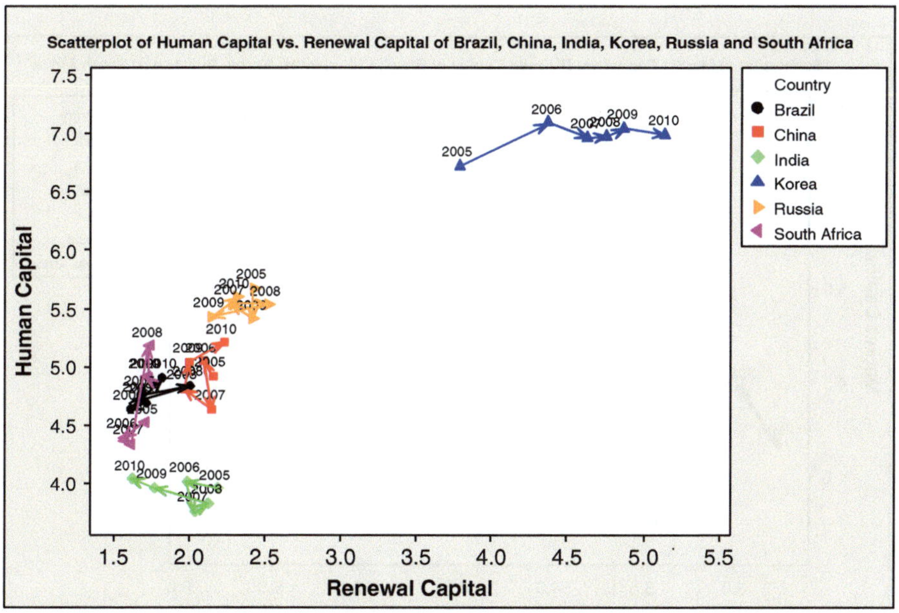

Fig. 3.13 Scatterplot of human capital vs. renewal capital for the BRICKS countries

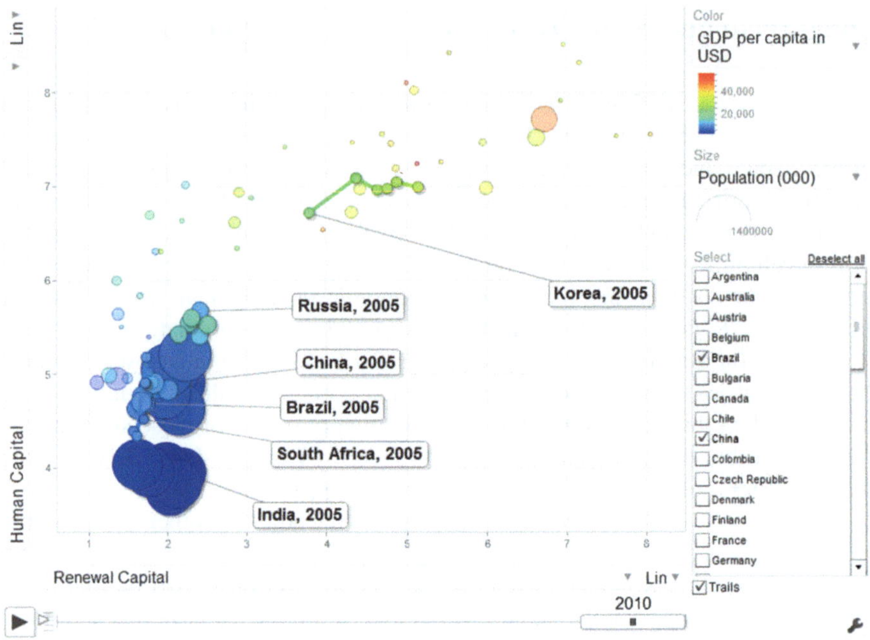

Fig. 3.14 Human capital vs. renewal capital for BRICKS countries

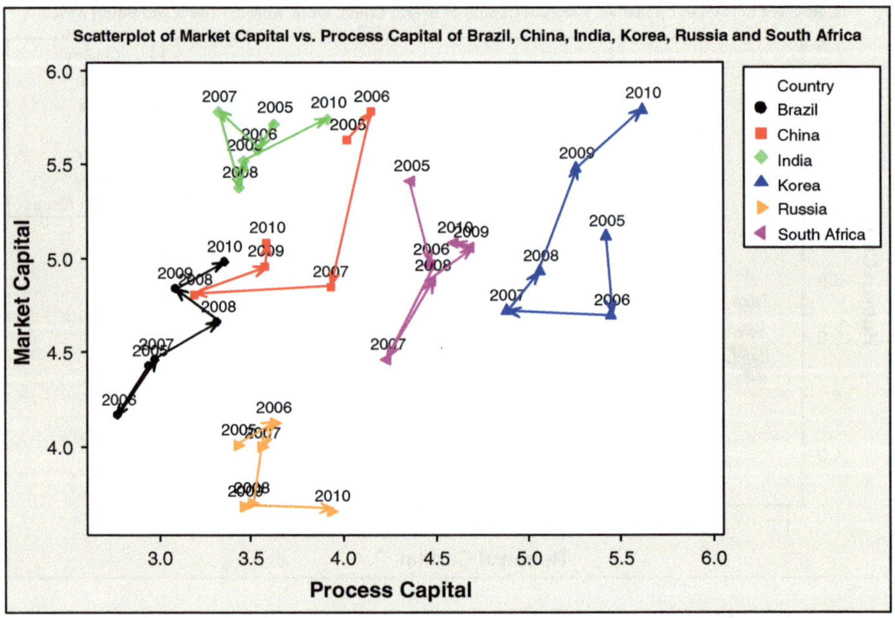

Fig. 3.15 Scatterplot of market capital vs. process capital for BRICKS countries

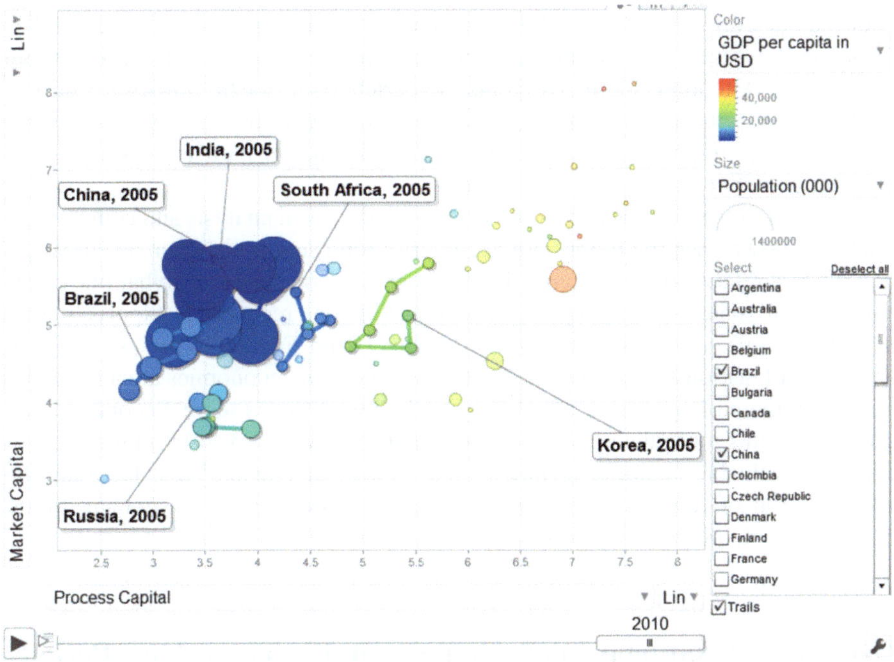

Fig. 3.16 Market capital vs. process capital for the BRICKS countries

capital, yet little improvement in renewal capital. Particularly, India was lagging behind in human capital. Since human capital and renewal capital take time to develop and achieve, such as skilled labor, R&D investments, patents and scientific articles, countries that are devoted to these two long-term NIC will possess the key factors for future sustainable competitiveness. When modeling the well-performing northern European countries, it is clear that they excel in these two long-term intellectual capitals. As the competition among the emerging economies will become more and more intense, these two graphs provide some angles for policy makers to ponder upon.

Figures 3.15 and 3.16 map market capital and process capital of the six countries on the same coordinate plane. Market capital refers to the general assets embodied in a nation's relationship with international markets while process capital refers to the national infrastructure, such as information systems and the number of days required to start a new business. In the short term, these are the two types of capital that can be expected to produce faster results if a nation is willing to engage in more extensive investments.

The following two short-term NIC graphs indicate that Korea is not too far apart from the rest of the countries. Russia's market capital is stagnant in more recent years even though its process capital progressed after the financial crisis. A warning sign for China is that its market capital and process capital were decreasing at a rapid pace before the financial crisis started. Even though China's short-term NIC bounced back after 2008, it is still lower than that of India's. In addition, Brazil has made good progress in both market capital and process capital and is catching up with China. Although South Africa also experienced a drastic downward turn in 2006 and 2007, its rebound after 2008 resulted in a higher process capital than that of China. This graph shows that China has been losing its short-term NIC edge to other emerging countries over the 6 years.

The 48 countries database we used is for closed system analysis and comparison. That is, a downward trend does not necessarily represent a real decrease in value, rather it is a relative decline comparing to other countries and somewhat like external competitiveness. For example, China showed strong recovery in exports (partial market capital), yet its market capital declined from Figs. 3.15 and 3.16. There are two explanations for this result. First, other emerging economies are also recovering with increasing exports, which may push down China's score. Second, market capital contains more than just objective statistics (such as export volume); it also embodies valuable "culture openness" and "image of country" that facilitate the accumulation of future market capital. Since NIC study aims at showing future potentiality, the presented results sound an early alarm for China.

Dynamics of National Intellectual Capital in Three Time Periods

The illustrations and graphs in the previous section describe NIC capital score changes in each country and the co-development path of various capitals with GDP. The ranking changes that will be introduced in this section represent external competitiveness. Rankings, by and large, is a zero-sum game in which the advance of one country simultaneously marks the retreat of another.

In order to probe the dynamics of NIC change before and after the financial crisis, we plotted Figs. 3.17 to 3.28 and Table 3.2 in three time periods: 2005–2006, 2007–2008, and 2009–2010. To have a clearer presentation, we separated the data into two sets. The first set contains capital scores and ranking changes of human capital, market capital, and process capital in Figs. 3.17, 3.19, 3.21, 3.23, 3.25, and 3.27. The second set exhibits capital scores and ranking changes of renewal capital, financial capital, and average NIC in Figs. 3.18, 3.20, 3.22, 3.24, 3.26, and 3.28.

The figures were then mapped based on the mean capital scores and ranking changes of each period. Ranking changes are next to each capital score and copied from the boldfaced ranking difference of columns 2, 5, 7, and 9 of Table 3.2. Taking the human capital ranking difference of Brazil (−1, 2, −3, −1) as an example, the ranking change of "−1," representing the difference between the third (2009–2010) and first (2005–2006) period, is plotted at 2005; the ranking change of "2," representing the difference between the second period (2007–2008) and first (2005–2006), is plotted at 2007; the ranking change of "−3," representing the difference between the third period (2009–2010) and second (2007–2008), is plotted at 2009; and the ranking change of "−1," representing the difference between the third period (2009–2010) and 6-year average ranking (exhibited in Table 3.1), is plotted at 2010.

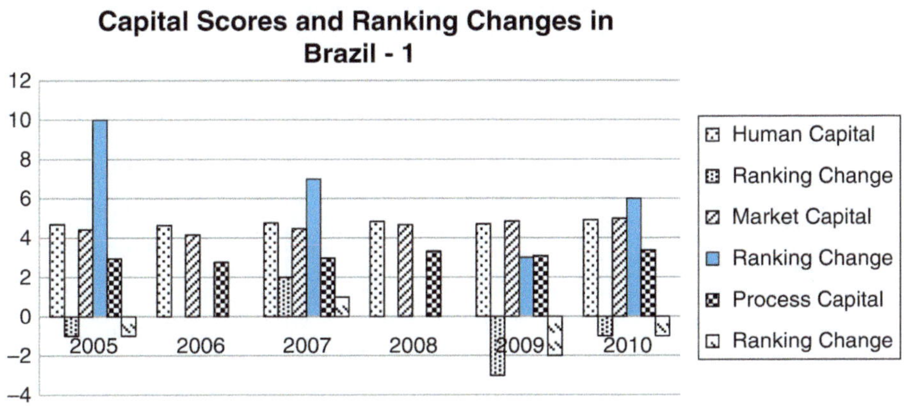

Fig. 3.17 Human capital, market capital, process capital, and ranking changes in Brazil

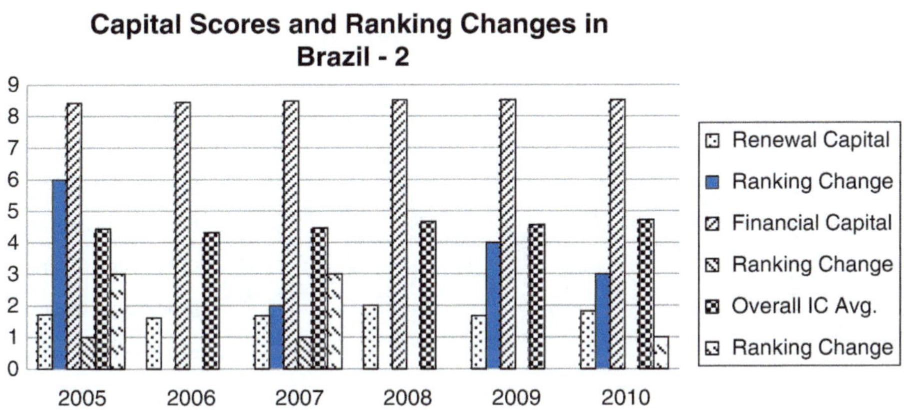

Fig. 3.18 Renewal capital, financial capital, average intellectual capital, and ranking changes in Brazil

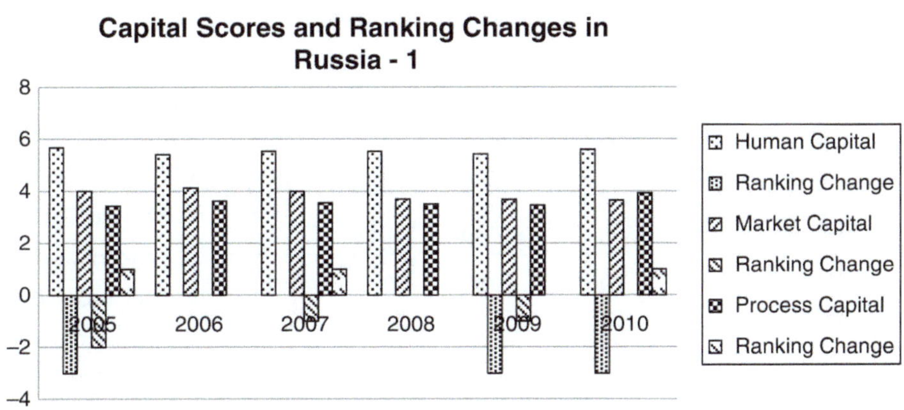

Fig. 3.19 Human capital, market capital, process capital, and ranking changes in Russia

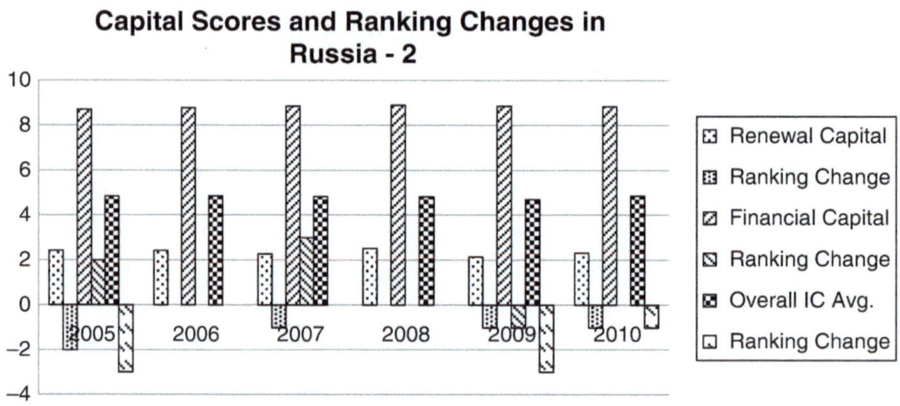

Fig. 3.20 Renewal capital, financial capital, average intellectual capital, and ranking changes in Russia

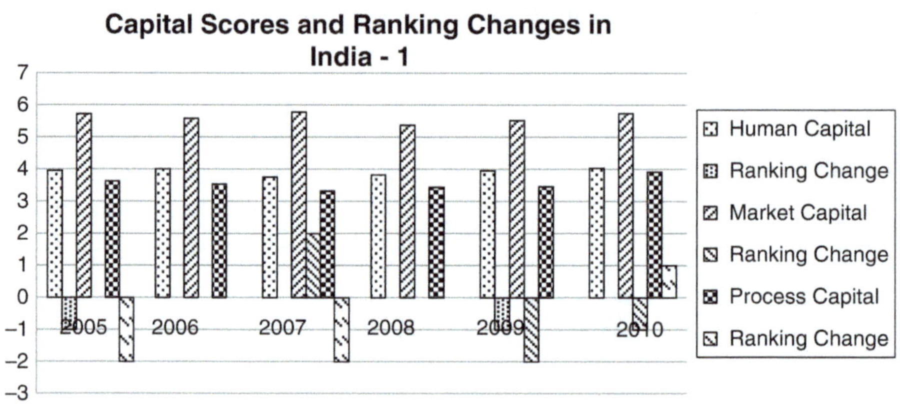

Fig. 3.21 Human capital, market capital, process capital, and ranking changes in India

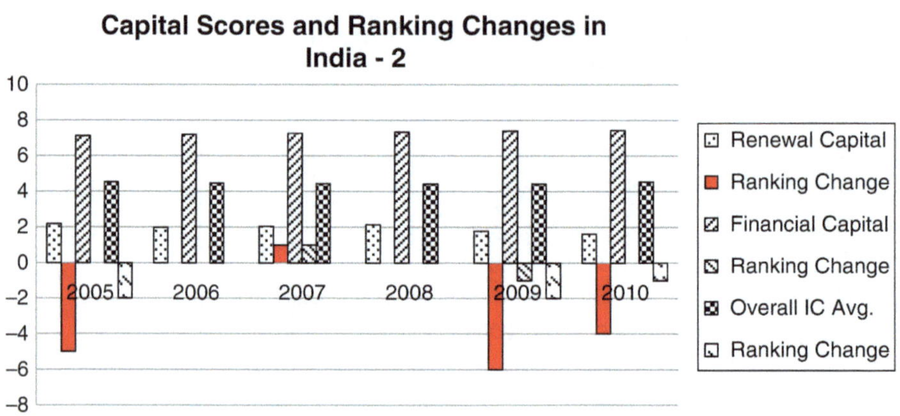

Fig. 3.22 Renewal capital, financial capital, average intellectual capital, and ranking changes in India

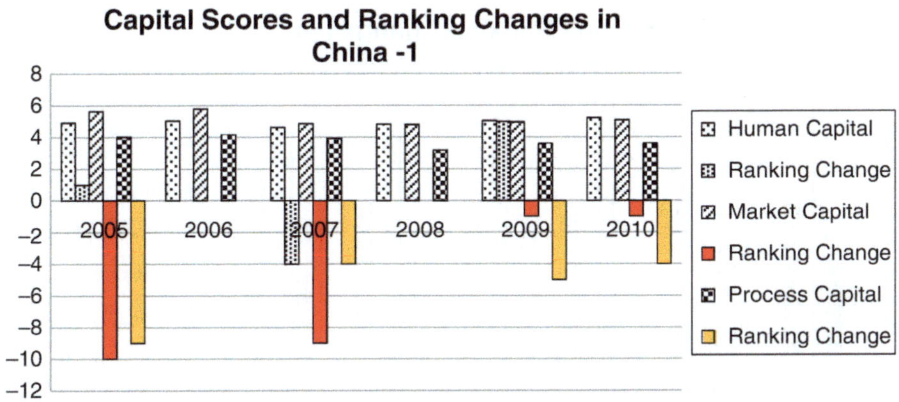

Fig. 3.23 Human capital, market capital, process capital, and ranking changes in China

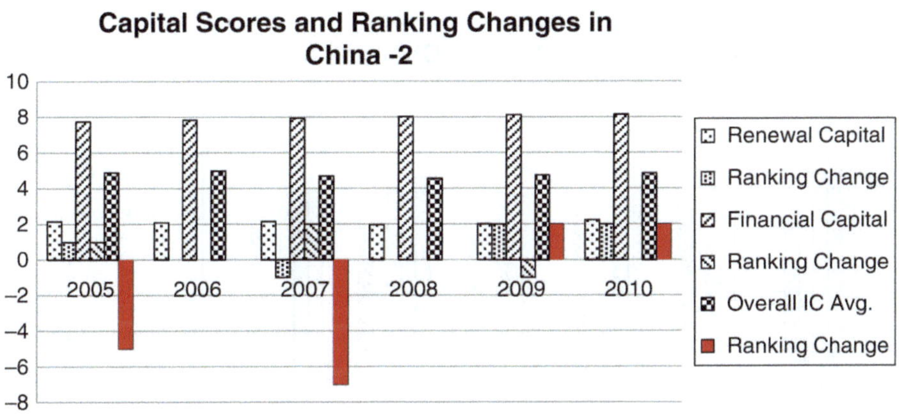

Fig. 3.24 Renewal capital, financial capital, average intellectual capital, and ranking changes in China

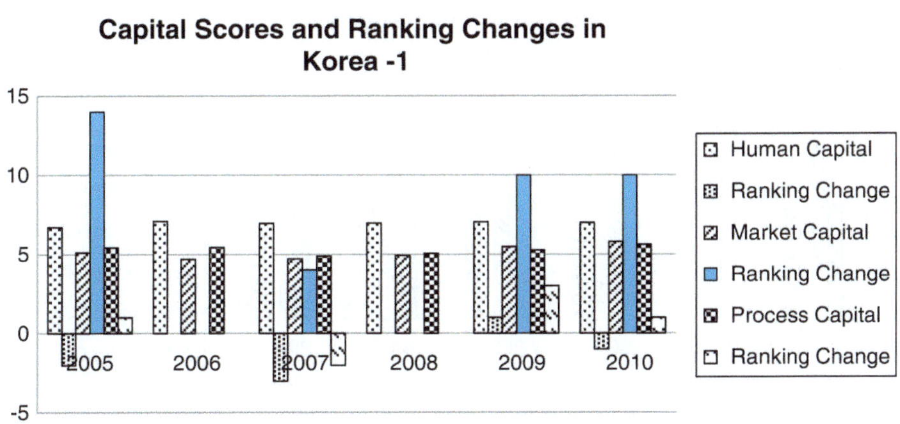

Fig. 3.25 Human capital, market capital, process capital, and ranking changes in Korea

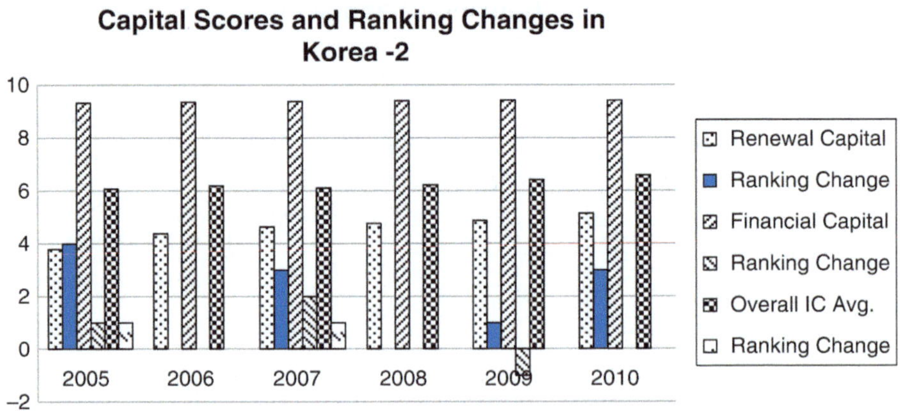

Fig. 3.26 Renewal capital, financial capital, average intellectual capital, and ranking changes in Korea

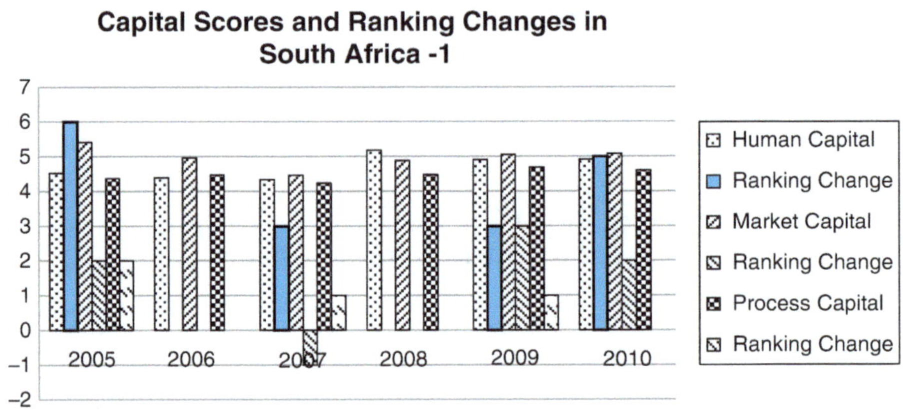

Fig. 3.27 Human capital, market capital, process capital, and ranking changes in South Africa

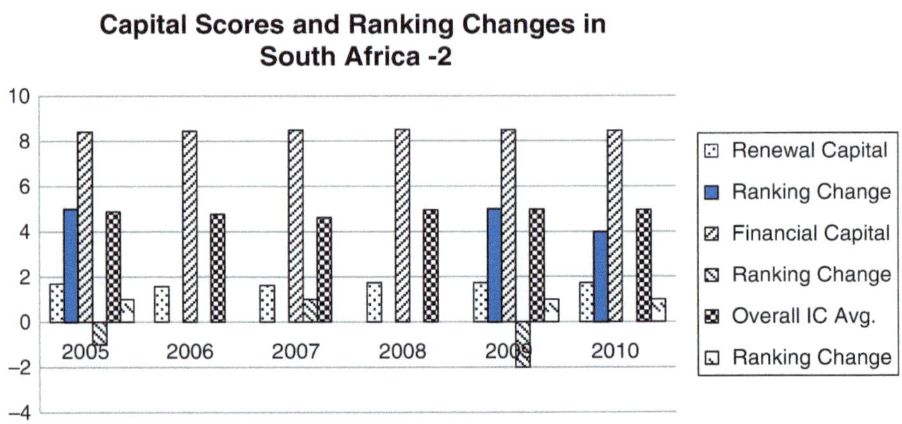

Fig. 3.28 Renewal capital, financial capital, average intellectual capital, and ranking changes in South Africa

Table 3.2 Ranking changes in three time periods for BRICKS countries

	(1st period to 3rd period)	1st period 2005–2006	2nd period 2007–2008		3rd period 2009–2010		3rd period ranking vs. overall NIC ranking	
Human capital	Difference#1	Ranking	Ranking	Difference	Ranking	Difference	NIC 48 Ranking	Difference
Brazil	**–1**	44	42	**2**	45	**–3**	44	**–1**
Russia	**–3**	33	33	**0**	36	**–3**	33	**–3**
India	**–1**	47	47	**0**	48	**–1**	48	**0**
China	**1**	40	44	**–4**	39	**5**	39	**0**
Korea	**–2**	18	21	**–3**	20	**1**	19	**–1**
South Africa	**6**	46	43	**3**	40	**3**	45	**5**
Market capital	Difference	Ranking	Ranking	Difference	Ranking	Difference	NIC 48 Ranking	Difference
Brazil	**10**	42	35	**7**	32	**3**	38	**6**
Russia	**–2**	44	45	**–1**	46	**–1**	46	**0**
India	**0**	21	19	**2**	21	**–2**	20	**–1**
China	**–10**	19	28	**–9**	29	**–1**	28	**–1**
Korea	**14**	33	29	**4**	19	**10**	29	**10**
South Africa	**2**	30	31	**–1**	28	**3**	30	**2**
Process capital	Difference	Ranking	Ranking	Difference	Ranking	Difference	NIC 48 Ranking	Difference
Brazil	**–1**	46	45	**1**	47	**–2**	46	**–1**
Russia	**1**	41	40	**1**	40	**0**	41	**1**
India	**–2**	39	41	**–2**	41	**0**	42	**1**
China	**–9**	35	39	**–4**	44	**–5**	40	**–4**
Korea	**1**	26	28	**–2**	25	**3**	26	**1**
South Africa	**2**	33	32	**1**	31	**1**	31	**0**

(continued)

Table 3.2 (continued)

Renewal capital	Difference	Ranking	Ranking	Difference	Ranking	Difference	NIC 48 Ranking	Difference
Brazil	**6**	39	37	**2**	33	**4**	36	**3**
Russia	**–2**	27	28	**–1**	29	**–1**	28	**–1**
India	**–5**	33	32	**1**	38	**–6**	34	**–4**
China	**1**	32	33	**–1**	31	**2**	33	**2**
Korea	**4**	20	17	**3**	16	**1**	19	**3**
South Africa	**5**	40	40	**0**	35	**5**	39	**4**
Financial capital	Difference	Ranking	Ranking	Difference	Ranking	Difference	NIC 48 Ranking	Difference
Brazil	**1**	41	40	**1**	40	**0**	40	**0**
Russia	**2**	33	30	**3**	31	**–1**	31	**0**
India	**0**	48	47	**1**	48	**–1**	48	**0**
China	**1**	45	43	**2**	44	**–1**	44	**0**
Korea	**1**	26	24	**2**	25	**–1**	25	**0**
South Africa	**–1**	40	39	**1**	41	**–2**	41	**0**
NIC	Difference	Ranking	Ranking	Difference	Ranking	Difference	NIC 48 Ranking	Difference
Brazil	**3**	44	41	**3**	41	**0**	42	**1**
Russia	**–3**	36	36	**0**	39	**–3**	38	**–1**
India	**–2**	43	43	**0**	45	**–2**	44	**–1**
China	**–5**	33	40	**–7**	38	**2**	40	**2**
Korea	**1**	23	22	**1**	22	**0**	22	**0**
South Africa	**1**	37	37	**0**	36	**1**	37	**1**

Remarks:

1. The second column compares the 3rd (2009–2010) period ranking with the 1st (2005–2006)
2. The fifth column compares the 2nd (2007–2008) period ranking with the 1st (2005–2006)
3. The seventh column compares the 3rd (2009–2010) period ranking with the 2nd (2007–2008)
4. The last column compares the most current 3rd period ranking (2009–2010) with the original 6-year average NIC48 ranking summarized in Table 3.1

Special attention can be paid to the negative numbers, which indicate that the level of a particular capital was higher before than after the financial crisis. Since it is easier to read figures than tables, we will describe the findings of each figure separately.

Figure 3.17 and Table 3.2 indicate Brazil's ranking changes in human capital (−1, 2, −3, −1), market capital (10, 7, 3, 6), and process capital (−1, 1, −2, −1) over the three time periods. Its relatively large ranking gains in market capital show that Brazil is gaining international competitiveness in market capital. On the contrary, its human capital and process capital have minor regress after the financial crisis.

Figure 3.18 shows all positive numbers, meaning Brazil advances its ranking in renewal capital (6, 2, 4, 3), financial capital (1, 1, 0, 0), and NIC (3, 3, 0, 1) over the three periods, when ranked in the 48 countries pool. In particular, the relatively large-scale ranking gains in renewal capital indicate its increasing renewal capital international competitiveness, casting a potential prospect.

Figure 3.19 and Table 3.2 indicate Russia's ranking changes in human capital (−3, 0, −3, −3), market capital (−2, −1, −1, 0), and process capital (1, 1, 0, 1) over the three time periods. Its ranking declines in human capital, particularly, send a warning sign. Russia used to boast of its high quality human resources, the ranking loss shows its international competitiveness in human capital is gradually losing ground.

Figure 3.20 shows that Russia's renewal capital (−2, −1, −1, −1) is being lowered by other countries for the four types of comparison, which may be associated with its regressing human capital ranking, as these two capitals are closely related. Russia's financial capital (2, 3, −1, 0) has some ranking gains after the financial crisis and its NIC (−3, 0, −3, −1) also lost international competitiveness after the financial crisis.

Figure 3.21 and Table 3.2 indicate India's ranking changes in human capital (−1, 0, −1, 0), market capital (0, 2, −2, −1), and process capital (−2, −2, 0, 1) over the three time periods. Although there are some negative numbers, the most recent comparison with the 6-year average ranking is about the same. However, India's strength in market capital had minor ranking declines and deserves some attention.

Figure 3.22 indicates that India has the greatest ranking decline in renewal capital (−5, 1, −6, −4). Its ranking changes in financial capital (0, 1, −1, 0) and NIC (−2, 0, −2, −1) are relatively minor. For a quick comparison, unlike India, Brazil advanced its renewal capital ranking on a relatively large scale, ranging from 2 to 6 as exhibited in Fig. 3.18. In future national development, India is at a disadvantage with lower renewal capital among emerging countries.

Figure 3.23 and Table 3.2 indicate China's ranking changes in human capital (1, −4, 5, 0), market capital (−10, −9, −1, −1), and process capital (−9, −4, −5, −4) over the three time periods. Its relatively large-scale ranking decline of the first two comparisons (−10, −9) and (−9, −4) in market capital and process capital explains that China was very strong in these two capitals in 2005–2006. When the financial crisis hit, they dropped drastically. Afterwards, China had a quick rebound in market capital, leading to a minor lost of one rank (−1) when comparing the most recent 2009–2010 with its 6-year average ranking. However, its rebound in process capital was not as good and still declined by four ranks (−4) with the same comparison. The development of China's process capital ranking changes deserves a special attention. Figure 3.3 shows that its process capital score ranges from 3.19 to 4.15. With about one point difference on a 1–10 scale, China was losing its international competitiveness in process capital on a relatively large scale.

Figure 3.24 shows the ranking changes of China's renewal capital (1, −1, 2, 2), financial capital (1, 2, −1, 0), and NIC (−5, −7, 2, 2). Although there are some negative numbers, the last comparisons are all positive, indicating China has rebounded in these capitals when comparing the most recent 2009–2010 with its 6-year average ranking.

Figure 3.25 and Table 3.2 indicate Korea's ranking changes in human capital (−2, −3, 1, −1), market capital (14, 4, 10, 10), and process capital (1, −2, 3, 1) over the three time periods. Its large-scale ranking gains in market capital indicates Korea gained international competitiveness in market capital after the financial crisis. Other negative numbers are relatively minor.

Figure 3.26 shows the ranking changes of renewal capital (4, 3, 1, 3), financial capital (1, 2, −1, 0), and NIC (1, 1, 0, 0) in Korea. Its renewal capital has ranking gains for all four comparisons, indicating Korea's potential in future-oriented and innovation-oriented capabilities. These ranking changes agree with Korea's demonstrated resilience during this financial crisis.

Figure 3.27 and Table 3.2 indicate South Africa's ranking changes in human capital (6, 3, 3, 5), market capital (2, −1, 3, 2), and process capital (2, 1, 1, 0) over the three time periods. Its relatively large-scale ranking gains in human capital indicates the country's future potential, as human capital is the very basis of national development. Other ranking changes with only one negative number are relatively minor.

Figure 3.28 shows the ranking changes of renewal capital (5, 0, 5, 4), financial capital (−1, 1, −2, 0), and NIC (1, 0, 1, 1) for South Africa. Its renewal capital also has relatively large-scale ranking gains, indicating its potential in future-oriented and innovation-oriented capabilities. Combining gaining international competitiveness in human capital and renewal capital (long-term NIC), South Africa is in good standing after the financial crisis to speed up its future national development.

In summary, for human capital ranking, the six BRICKS countries have ranking gains as well as ranking declines. The ranking decline in Russia is particularly worrying, as it used to boast good quality human resources. The relatively large-scale ranking gains in South Africa is a promising sign for the country's future development. For market capital, the large-scale ranking gains of Brazil and Korea for all four comparisons are particularly eye-catching, indicating their good recovery after the financial crisis. On the contrary, China's largest scale ranking decline in the first two comparisons (−10, −9) provides a point of concern. For process capital, China has the greatest ranking declines among the six countries.

For renewal capital, the relatively large-scale ranking gains of Brazil, Korea, and South Africa are in contrast with India's ranking lost. For financial capital, the ranking variation of this country cluster is relatively small. For NIC, although there are ranking declines, most of the countries regained their 6-year average ranking. Particularly, China gained two ranks in NIC after the financial crisis, even though it had gone through drastic ranking drops in market capital and process capital.

3-Dimensional National Intellectual Capital Trajectory

Up until this section, all the figures plotted different combinations of various types of capital, capital-GDP co-development, and ranking changes on a two-dimensional plane. Even though Figs. 3.7, 3.8, 3.9, 3.10, 3.11, and 3.12 displayed four dimensions with color and bubble size as the third and fourth dimensions in addition to *X* axis and *Y* axis, graphically they are still on one plane. This section presents three-dimensional (3D) graphs showing each country's NIC development trajectory on a 48-country 3D landscape of various types. Since NIC evolution over time is always complex, combining different views from different angles may provide more comprehensive perspectives for analysis.

In what follows, we first explore the value driving context by uncovering the path of GDP growth for Brazil, Russia, India, China, Korea, and South Africa from 2005 to 2010. Second, in explaining the GDP growth, we present the high, middle, and low capabilities which human capital, market capital, process capital, and renewal capital, respectively, have. Third, we examine the efficiency (GDP growth) enhancing factors and impeding factors of each country, taking each country's turning point into consideration. Last, we identify the efficiency drivers of each country in reaching a targeted GDP.

Figure 3.29 shows a rectangular 3D landscape, with a longer north–south side. The formation of the landscape is based on the data of 29 national intellectual

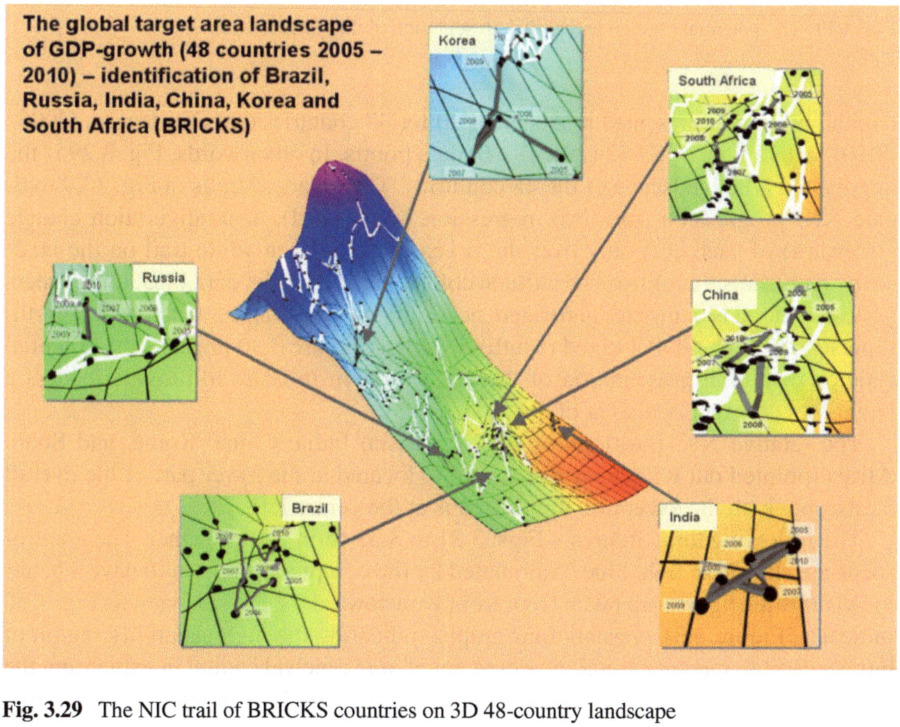

Fig. 3.29 The NIC trail of BRICKS countries on 3D 48-country landscape

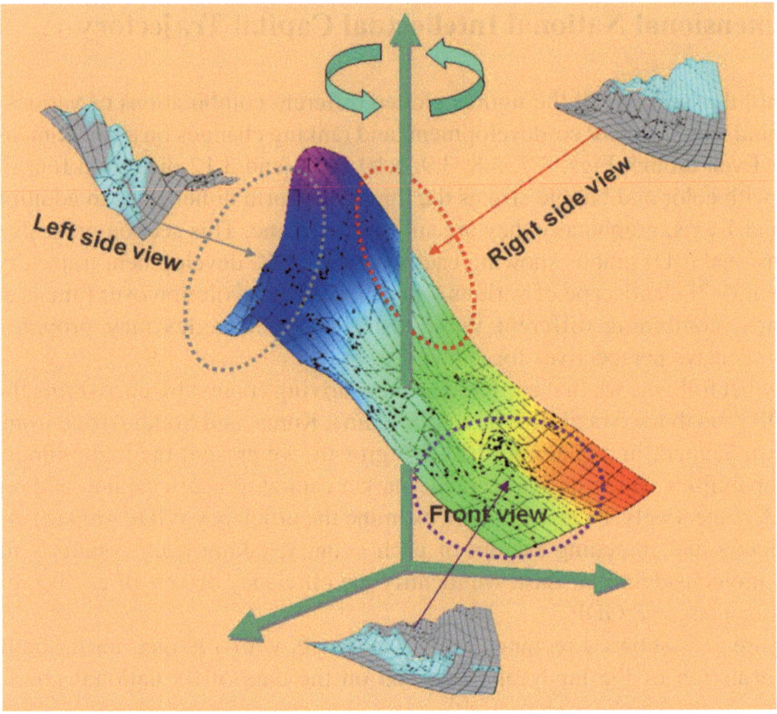

Fig. 3.30 The potential rotation and partial presentation of the 3D formation

capital indicators (presented in Appendix 3) for 48 countries covering 6 years (2005–2010), with a total of 8,352 ($29 \times 48 \times 6$) data points. In other words, Fig. 3.29 is the aggregated NIC landscape of the 48 countries. The enlarged trails in Fig. 3.29 indicate NIC progression (upward), regression (downward), or configuration change (sideward) of each economy over the 6-year period. Each white trail on the landscape represents a country. The altitude coloring from purple, dark blue, blue, green, yellow, and red is computer generated, according to GDP values, with purple (north) representing high GDP and red (south) low GDP. Figure 3.30 presents the potential rotation or partial presentation of the 3D formation in order to provide readers a better view for the focal area of discussion.

The relative NIC positions of Brazil, Russia, India, China, Korea, and South Africa (pointed out with arrows) are mainly located at the lower part of the overall landscape, with the exception that Korea is at the center left.

The following three figures—Figs. 3.31 to 3.33 form a set together. The areas of focus are shaded in light blue. Automated by the computer program to have a better view, the three figures are taken from front view toward the left side view as Fig. 3.30 indicates. Figure 3.31 presents four graphs indicating the high capability region of human capital, market capital, process capital, and renewal capital in explaining the GDP growth. High capability means the data points are 20% above the mean value

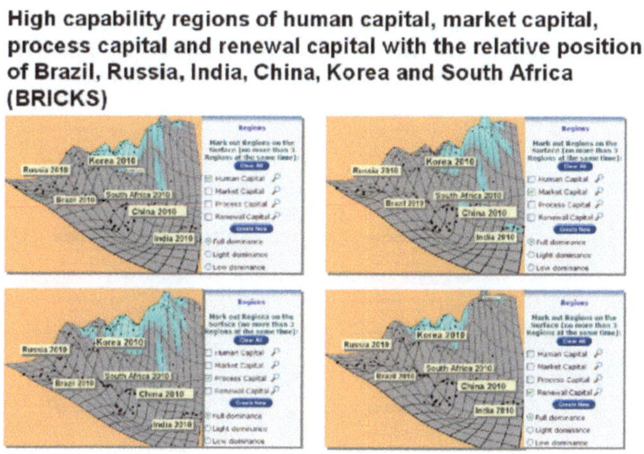

Fig. 3.31 The High capability region of human capital, market capital, process capital, and renewal capital

Fig. 3.32 The middle capability region of human capital, market capital, process capital, and renewal capital

of each capital. This figure shows that no capital in all six countries is in the high capability region.

Figure 3.32 displays another four graphs indicating the middle capability region of each human capital, market capital, process capital, and renewal capital in explaining the GDP growth. Middle capability means the data points are within the top 50% value of each capital. This figure shows that market capital of China and India and all four capitals of Korea is in the middle capability region.

Figure 3.33 presents the last four graphs in this set and indicates the low capability region of human capital, market capital, process capital, and renewal capital in explaining the GDP growth. Low capability means the data points are 20% below

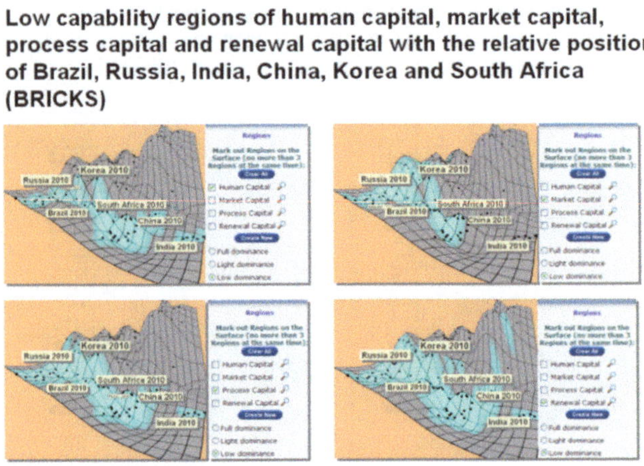

Fig. 3.33 The low capability region of human capital, market capital, process capital, and renewal capital

the mean value of each capital. This figure shows all the capitals other than what described in Fig. 3.32 are in the low capability region.

In general, Fig. 3.31 to 3.33 as a set explain that almost 75% of the capitals (18 out of $4 \times 6 = 24$) of these six countries are in the low capability region in explaining GDP growth. Even though China and India are at the bottom tier of NIC (ranked #40 and #44 out of 48), their market capital is in the middle capability region. These three figures demonstrate that the NIC of the BRICKS countries is waiting for further development.

Another set of figures, Fig. 3.34 to 3.39 display efficiency enhancing or impeding factors that affect GDP growth before and after the NIC turning point for BRICKS countries. The white indicator boxes at the upper right display the results of the enhancing or impeding factors. We employed three or four rounds of data analysis, depending on whether the country has one or multiple NIC turning points. For instance, Russia's turning points are 2008 and 2009; therefore, we trace the explaining factors of 2005 against 2008 (starting point → the first turning point), 2007 against 2008 (one year before the turning point → the first turning point), 2008 against 2009 (one year before the turning point → the second turning point), and 2009 against 2010 (the second turning point → the ending point).

As an easy reference, we summarize in Table 3.3 two sets of results, each with the five key factors. If there is a clear percentage difference cutting point (such as less than 1% in the white indicator boxes), the reported factors may be less than five. Results for the first set show the efficiency enhancing and impeding factors before the turning point. Results for the second set show the efficiency enhancing and impeding factors after the (last) turning point in italic letters. Since the factor strength (percentage) varies with different explaining power, we have calculated the average weight of the five enhancing and impeding factors and displayed each respective percentage in the rows for "Factor #5." We also count the frequency of

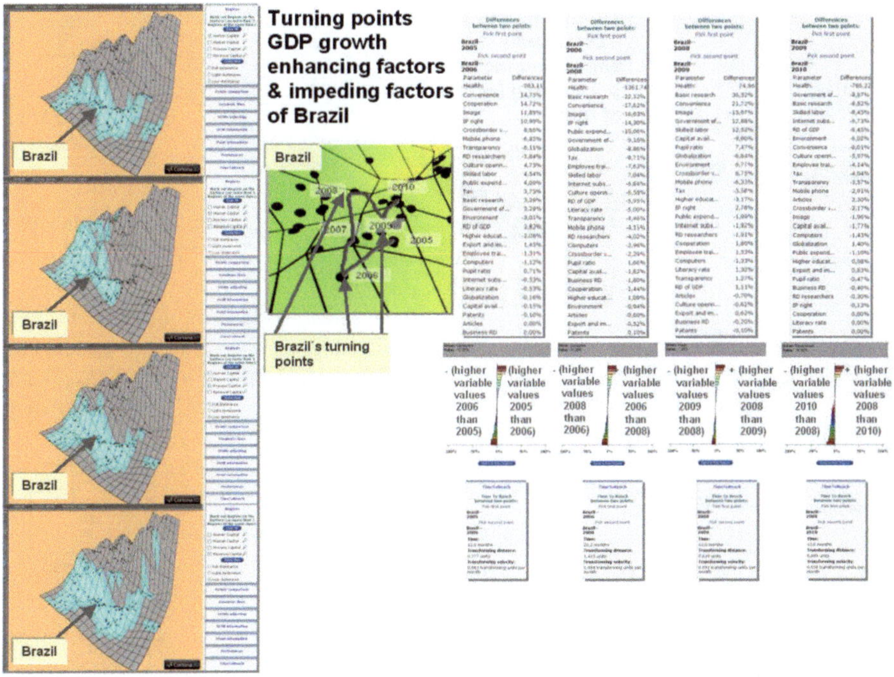

Fig. 3.34 Turning points and GDP growth enhancing and impeding factors of Brazil

Fig. 3.35 Turning points and GDP growth enhancing and impeding factors of Russia

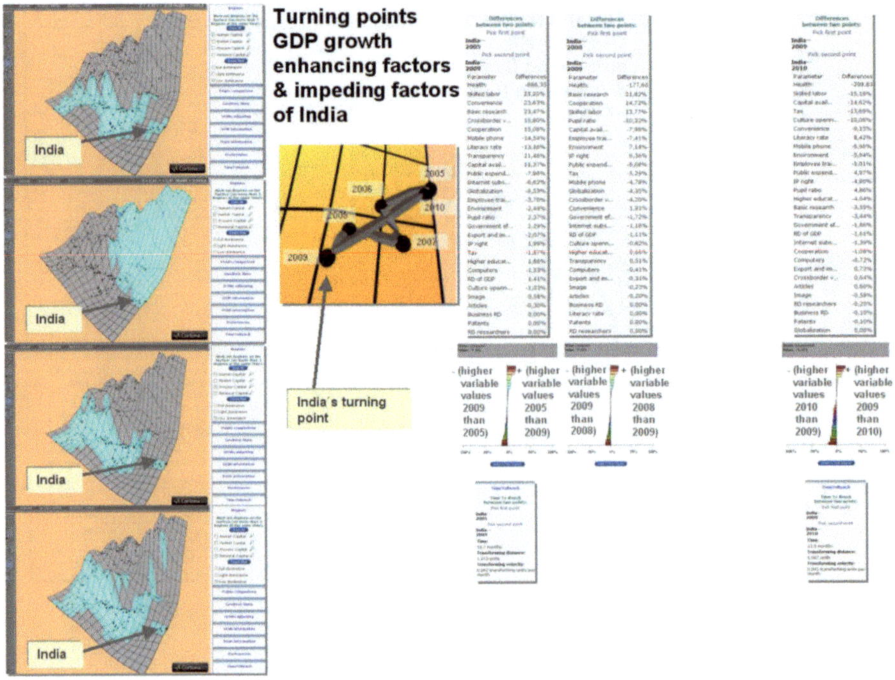

Fig. 3.36 Turning point and GDP growth enhancing and impeding factors of China

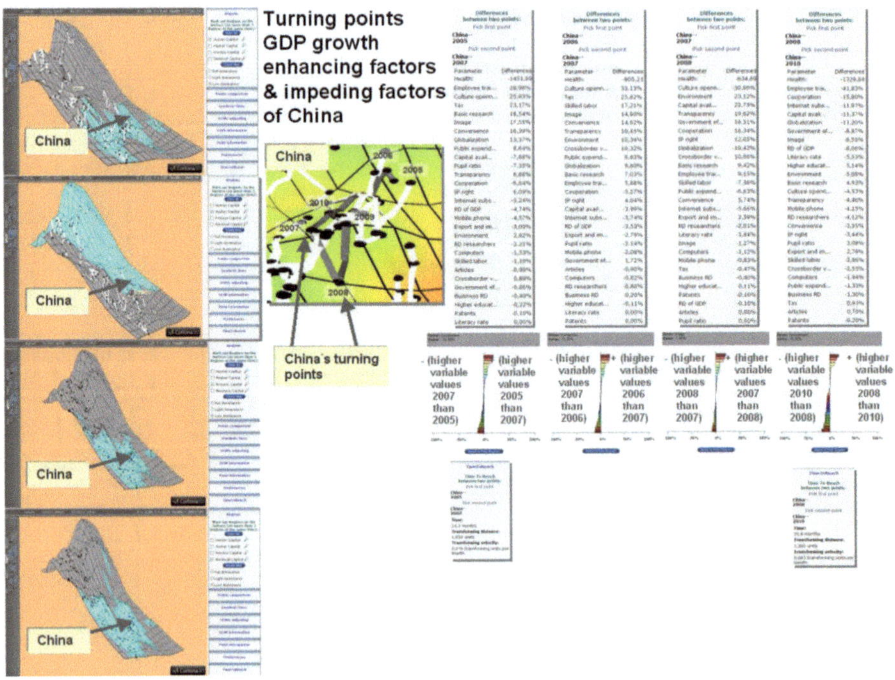

Fig. 3.37 Turning points and GDP growth enhancing and impeding factors of India

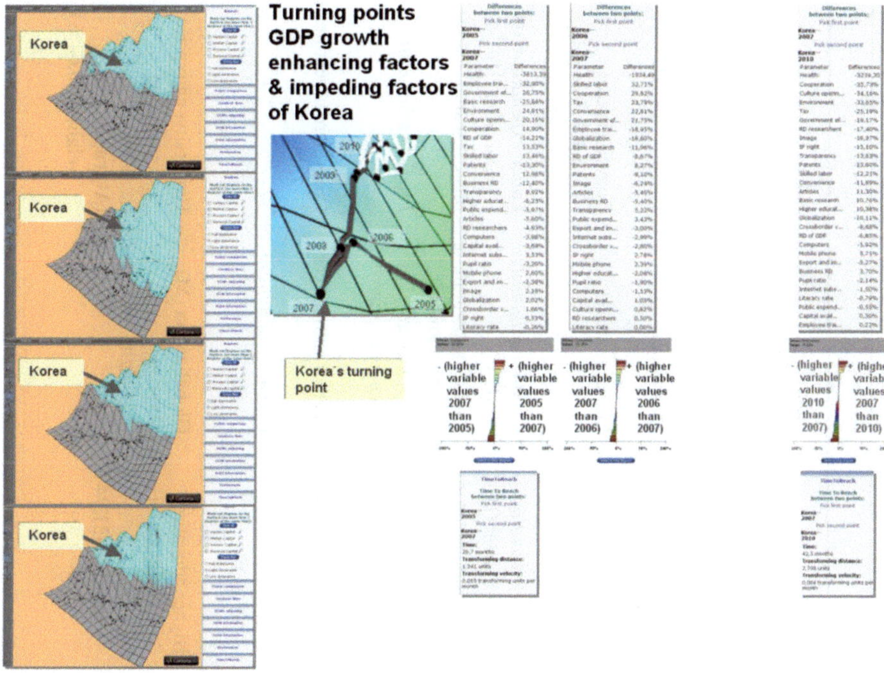

Fig. 3.38 Turning point and GDP growth enhancing and impeding factors of Korea

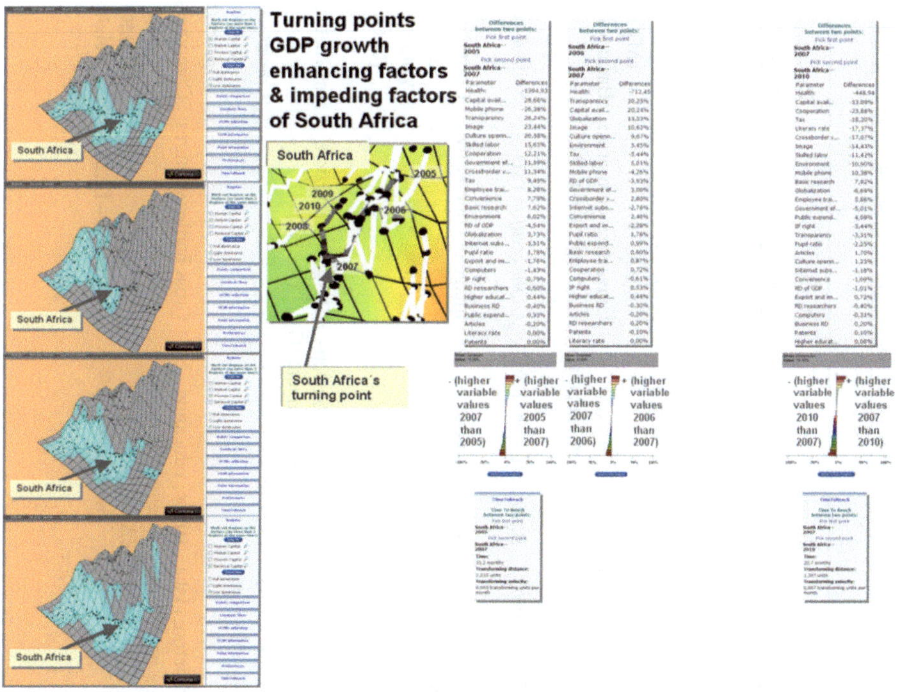

Fig. 3.39 Turning point and GDP growth enhancing and impeding factors of South Africa

Table 3.3 Enhancing factors and impeding factors of GDP per capita (ppp) growth for the six BRICKS countries

(a)

Influencing factor	Brazil Enhancing factors	Brazil Impeding factors	Russia Enhancing factors	Russia Impeding factors	India Enhancing factors	India Impeding factors
Factor #1 before (1st) turning point	Mobile phone subscribers	Convenience of establishing new firms (G)	Mobile phone subscribers	Capital availability (G-C)	Mobile phone subscribers	Skilled labor (H)
Factor#1 after (last) turning point	*Government efficiency* (G)	*Scientific articles* (R&D)	*Openness of culture* (M)	*Corporate tax encouragement* (G)	*Skilled labor* (H)	*Literacy rate* (H)
Factor #2 before (1st) turning point	Transparency of government policy (G)	University–enterprise cooperation (R&D)	Basic research (R&D)	Employee training (H)	Literacy rate (H)	Convenience of establishing new firms (G)
Factor #2 after (last) turning point	*Basic research* (R&D)	*Globalization* (M)	*Skilled labor* (H)	*Higher education enrollment* (H)	*Capital availability* (G-C)	*Public expenditure on education* (H)
Factor #3 before (1st) turning point	R&D researchers (R&D)	Image of country (M)	Corporate tax encouragement (G)	University–enterprise cooperation (R&D)	Public expenditure on education (H)	Basic research (R&D)
Factor #3 after (last) turning point	*Skilled labor* (H)	–	*Capital availability* (G-C)	*Employee training* (H)	*Corporate tax encouragement* (G)	*Pupil–teacher ratio* (H)
Factor #4 before (1st) turning point	Fair business competition environment (G)	IP rights protection (R&D)	Internet subscribers	Openness of culture (M)	Internet subscribers	Cross-border ventures (M)
Factor #4 after (last) turning point	*Internet subscribers*	–	*Convenience of establishing new firms* (G)	*Scientific articles* (R&D)	*Openness of culture* (M)	–
Factor #5 before (1st) turning point	Higher education enrollment (H) **(Average weight—4.78%)**[a]	Cross border ventures (M) **(Average weight—12.20%)**	Computers in use per capita **(Average weight—17.64%)**	Higher education enrollment (H) **(Average weight—21.94%)**	Globalization (M) **(Average weight—9.76%)**	University–enterprise cooperation (R&D) **(Average weight—20.64%)**

Factor #5 after (last) turning point	*R&D % GDP* (R&D) **(Average weight—7.86%)**	**(Average weight—1.85%)**	*Basic research* (R&D) **(Average weight—11.95%)**	*Pupil–teacher ratio* (H) **(Average weight—8.23%)**	*Convenience of establishing new firms* (G) **(Average weight—12.54%)**	– **(Average weight—6.08%)**
Category frequency count	G – 3 H – 2 M – 0 R&D – 3	G –1 H – 0 M – 3 R&D – 3	G – 3 H – 1 M – 1 R&D – 2	G – 2 H – 5 M – 1 R&D – 2	G – 3 H – 3 M – 2 R&D – 0	G – 1 H – 4 M – 1 R&D – 2

(continued)

Table 3.3 (continued)

(b)

Influencing factor	China Enhancing factors	China Impeding factors	Korea Enhancing factors	Korea Impeding factors	South Africa Enhancing factors	South Africa Impeding factors
Factor #1 before (1st) turning point	Capability availability (G-C)	Employee training (H)	Employee training (H)	Government efficiency (G)	Mobile phone subscribers	Capital availability (G-C)
Factor#1 after (last) turning point	*Employee training* (H)	*Higher education enrollment* (H)	*University–enterprise cooperation* (R&D)	*Scientific articles* (R&D)	*Capital availability* (G-C)	*Mobile phone subscribers*
Factor #2 before (1st) turning point	Pupil–teacher ratio (H)	Openness of culture (M)	Basic research (R&D)	Fair business competition environment (G)	R&D % of GDP (R&D)	Transparency of government policy (G)
Factor #2 after (last) turning point	*University–enterprise cooperation* (R&D)	*Basic research* (R&D)	*Openness of culture* (M)	*Basic research* (R&D)	*Enterprise–university cooperation* (R&D)	*Basic research* (R&D)
Factor #3 before (1st) turning point	University–enterprise cooperation (R&D)	Corporate tax encouragement (G)	R&D % GDP (R&D)	Openness of culture (M)	Internet subscribers	Image of country (M)
Factor #3 after (last) turning point	*Internet subscribers*	*Pupil–teacher ratio* (H)	*Fair business competition environment* (G)	*Higher education enrollment* (H)	*Corporate tax encouragement* (G)	*Employee training* (H)
Factor #4 before (1st) turning point	Internet subscribers	Basic research (R&D)	Patents (R&D)	University–enterprise cooperation (R&D)	Export of goods (M)	Openness of culture (M)
Factor #4 after (last) turning point	*Capital availability* (G-C)	*Export of goods* (M)	*Corporate tax encouragement* (G)	–	*Literacy rate* (H)	*Public expenditure on education* (H)
Factor #5 before (1st) turning point	R&D % of GDP (R&D) **(Average weight—6.33%)**[a]	Image of country (M) **(Average weight—22.83%)**	Business R&D (R&D) **(Average weight—19.73%)**	Corporation tax encouragement (G) **(Average weight—20.03%)**	Computers in use per capita **(Average weight—7.48%)**	Skilled labor (H) **(Average weight—22.91%)**
Factor #5 after (last) turning point	*Globalization* (M) **(Average weight—18.431%)**	– **(Average weight—3.99%)**	*Government efficiency* (G) **(Average weight—29.58%)**	– **(Average weight—10.81%)**	*Cross border ventures* (M) **(Average weight—21.92%)**	*Scientific articles* (R&D) **(Average weight—5.99%)**

Category frequency count	G – 2 H – 2 M – 1 R&D – 3	G – 1 H – 3 M –3 R&D – 2	G – 3 H – 1 M – 1 R&D – 5	G –3 H – 1 M – 1 R&D – 3	G – 2 H – 1 M – 2 R&D – 2	G – 2 H – 3 M – 2 R&D – 2
Category frequency count – **Grand total**	G –16 H – 10 M – 7 R&D – 15	G – 10 H – 16 M –11 R&D – 14				

Remarks: (G & G-C)—government-related issues including capital availability represented by G-C (total 26 counts including 7 counts of capital availability), (H)—human capital issues (26 counts), (M)—market capital issues (18 counts), (R&D)—research and development (29 counts)

[a]Average explaining power of each enhancing/impeding factor before and after the turning point

four more frequently displayed categories, namely government-related issues (G), human capital-related issues (H), market capital-related issues (M), and research and development issues (R&D). Since capital availability was a concern during the financial crisis, it was particularly highlighted as G-C.

Figure 3.34 shows the 2006, 2008, and 2009 turning points and the enhancing and impeding factors of Brazil. The four rectangular 3D graphs on the left-hand side of the figure indicate that all of Brazil's human capital, market capital, process capital, and renewal capital fall into the low capability region (20% below average).

Table 3.3 shows that in Brazil, the major enhancing factors include government and R&D-related issues of 3 counts each, however with relatively low average weight of 4.78% and 7.86% before and after the turning point, respectively. On the other hand, the major impeding factors include market capital- and R&D-related issues also 3 counts each. The average weight of 12.20% before the 2006 turning point indicates that the five impeding factors, namely convenience of establishing new firms, university–enterprise cooperation, image of country, IP rights protection, and cross-border ventures deserve some attention. After the 2009 turning point, there are only two impeding factors with a very light average weight of 1.85%.

Figure 3.35 shows the 2008 and 2009 turning points and the enhancing and impeding factors of Russia. The four rectangular 3D graphs on the left-hand side of the figure indicate that all of Russia's human capital, market capital, process capital, and renewal capital fall into the low capability region (20% below average).

Table 3.3 shows that in Russia, the major enhancing factors include 3 counts of government and 2 counts of R&D-related issues, with a higher average weight before than after the turning point (17.64% vs. 11.95%). However, one government-related issue—corporate tax encouragement was turned from an enhancing factor before the turning point to impeding factor after the turning point. On the other hand, the major impeding factors include 5 counts of human capital and 2 counts each of government- and R&D-related issues. The average weight of 21.94% before the 2008 turning point indicates that the five impeding factors, namely capital availability, employee training, university–enterprise cooperation, openness of culture, and higher education enrollment deserve some attention. After the 2009 turning point, the average weight of 8.23% indicates that the stress of impeding factors was lessened. However, employee training was an impeding factor before and after the turning point. There are two good signs in that capital availability and openness of culture were turned from impeding factors before the turning point to enhancing factors after the turning point.

Figure 3.36 shows the 2009 turning point and the enhancing and impeding factors of India. The four rectangular 3D graphs on the left-hand side of the figure indicate that India's human capital, process capital, and renewal capital fall into the low capability region (20% below average). However its market capital is in the middle capability region (top 50%).

Table 3.3 shows that in India, the major enhancing factors include 3 counts each of government- and human capital-related issues, with a higher average weight after (12.54%) than before (9.76%) the turning point. However, 2 counts of human capital issues—literacy rate and public expenditure on education were turned from an enhancing factor before the turning point to impeding factor after the turning point, very likely due to the impact of financial crisis. On the other hand, the major impeding factors include 4 counts of human capital- and 2 counts of R&D-related issues. The average weight of 20.64% before the 2009 turning point indicates that the five impeding factors, namely skilled labor, convenience of establishing new firms, basic research, cross-border venture, and university–enterprise cooperation deserve some attention. After the 2009 turning point, the average weight of 6.08% indicates that the stress of impeding factors was lessened. However, convenience of establishing new firms as an impeding factor before the turning point was turned into an enhancing factor after the turning point.

Figure 3.37 shows the 2007 and 2008 turning points and the enhancing and impeding factors of China. The four rectangular 3D graphs on the left-hand side of the figure indicate that China's human capital, process capital, and renewal capital fall into the low capability region (20% below average); however, its market capital is in the middle capability region (top 50%).

Table 3.3 shows that in China, the major enhancing factors include 3 counts of R&D and 2 counts each of government- and human capital-related issues, with a higher average weight after (18.43%) than before (6.33%) the turning point. Capital availability, university–enterprise cooperation, and Internet subscribers remain as enhancing factors before and after the turning point. Particularly, the capital availability as a continued enhancing factor shows that capital is not a problem for China before and after the financial crisis. However, pupil–teacher ratio was turned from an enhancing factor before the turning point to impeding factor after the turning point. On the other hand, the major impeding factors include 3 counts each of human capital- and market capital-related issues. The average weight of 22.83% before the 2007 turning point is much higher than that (3.99%) after the 2008 turning point. Therefore, employee training, openness of culture, corporate tax encouragement, basic research, and image of country deserve some attention. However, employee training as an impeding factor before the turning point was turned to enhancing factor after the turning point.

Figure 3.38 shows the 2007 turning point and the enhancing and impeding factors of Korea. The four rectangular 3D graphs on the left-hand side of the figure indicate that all Korea's human capital, market capital, process capital, and renewal capital fall into the middle capability region (top 50%).

Table 3.3 shows that in Korea, the major enhancing factors include 5 counts of R&D and 3 counts of government-related issues, with a higher average weight after (29.58%) than before (19.73%) the turning point. The high average weight of five enhancing factors after the turning point include university–enterprise cooperation, openness of culture, fair business competition environment, corporate tax encouragement, and government efficiency. However, basic research as an enhancing factor before the turning point was turned to an impeding factor after the turning point. On the other hand, the major impeding factors include 3 counts each of government- and R&D-related issues. The average weight of 20.03% before the turning point is higher than that (10.81%) after the turning point. However, all the five impeding factors before the turning point were all turned to enhancing factor after the turning point, causing the high average weight (29.58%) of enhancing factor after the turning point. This finding is surprising in that the negative factors were turned to positives in such an efficient and complete way.

Figure 3.39 shows the 2007 turning point and the enhancing and impeding factors of South Africa. The four rectangular 3D graphs on the left-hand side of the figure indicate that all South Africa's human capital, market capital, process capital, and renewal capital fall into the low capability region (20% below average).

Table 3.3 shows that in South Africa, the major enhancing factors include 2 counts each of government-, market capital-, and R&D-related issues, with a higher average weight after (21.92%) than before (7.48%) the turning point. The five enhancing factors after the turning point include capital availability, university–enterprise cooperation, corporate tax encouragement, literacy rate, and cross-border ventures. On the other hand, the major impeding factors include 3 counts of human capital and 2 counts each of government-, market capital-, and R&D-related issues. The average weight of 22.91% before the turning point is much higher than that (5.99%) after the turning point. Therefore, capital availability, transparency of government policy, image of country, openness of culture, and skilled labor deserve some attention. However, capital availability as an impeding factor before the turning point was turned to enhancing factor after the turning point.

To summarize Table 3.3 into an overall suggestion, we count the frequency of both enhancing and impeding factors according to their categories for all the six countries, shown in the last row of the table. There are a total of 26 counts of government-related issues, including 7 counts of capital availability. Among these, there are 10 counts of impeding factors, with 2 counts of capital availability included. In other words, there are 5 counts of capital availability as an enhancing factor, with two for China, one each for India, Russia, and South Africa. This finding indicates the capital availability is not a big problem for the BRICKS countries in general.

There are 26 counts of human capital issues (H). Among which are 16 counts of impeding factors. The result is understandable as a majority of the BRICKS are emerging countries with relatively low human capital. There are 18 counts of market capital issues (M). Among these there are 11 counts of impeding factors. For research and development issues (R&D) there are 29 counts, with 14 counts of impeding factors. R&D issues turned out to have the greatest number of frequency count, even though some of them have relatively low weight. However, the finding did show a guideline for the next stage development of the BRICKS countries.

Figure 3.40 identifies the efficiency drivers (positive term) for each individual country. Efficiency targeting can be seen as a form of benchmarking. Efficiency represents the GDP in this 3D software program. Since Japan has been the model of Asian development in the past, it serves as a good benchmark. Therefore, Japan's GDP is the efficiency target in Fig. 3.40. In this figure, the stacked small horizontal stripes are the deficiency factors of each country benchmarked against Japan's GDP in 2010. The length of the stripes represents the degree of deficiency. For easier comparison, we list the first five deficiency factors for each economy in Table 3.4.

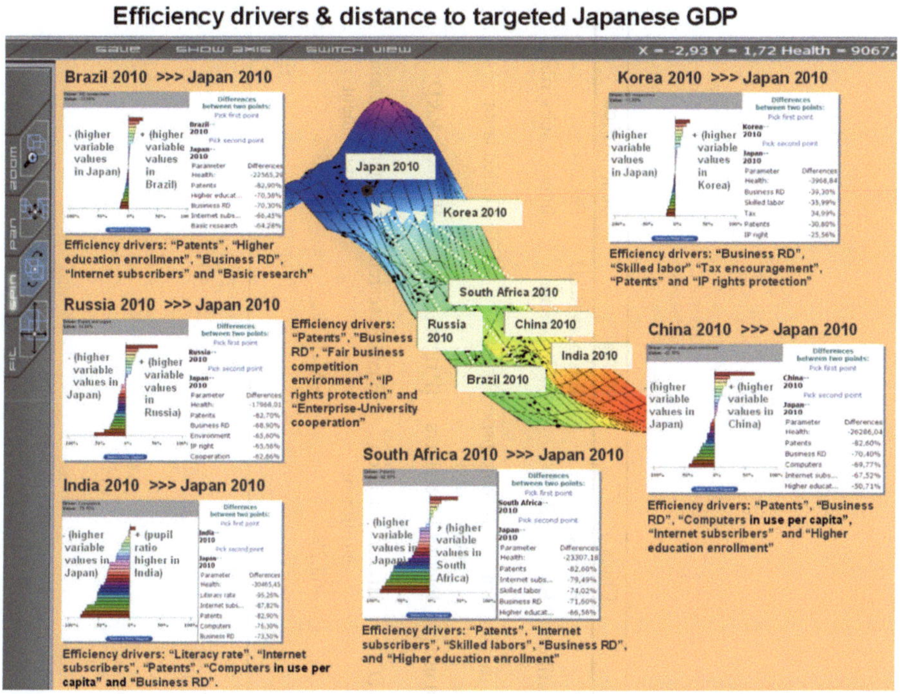

Fig. 3.40 Efficiency drivers and distance to targeted GDP of Japan

With Japan being used as a benchmarking country, irrespective of the sequence of deficiency, the common areas (in frequency count) that these six countries need to improve include patents (6 counts), business R&D spending (6 counts), Internet subscribers (4 counts), higher education enrollment (3 counts), and 2 counts each for basic research, IP rights protection, university–enterprise cooperation, and skilled labor.

From Table 3.4, 18 out of the 30 (5 × 6 countries) counts are R&D related, including 4 counts each for Korea and Russia, 3 counts each for Brazil and China, and 2 counts each for India and South Africa. To achieve the level of Japan's GDP in 2010, apparently enhancing R&D is very important. This result seems to reflect the level of economic development as well. For the relatively less-developed countries, India and South Africa, there are other issues that have an immediate concern. In addition to the two R&D issues, India needs to enhance its literacy rate, Internet subscribers, and computers in use per capita. South Africa has to enhance its Internet subscribers, skilled labor, and higher education enrollment.

Based on the average deficiency (negative term) percentage reported in Table 3.4, India has the longest distance (−83.16%) to achieve the target GDP (Japan's), followed by South Africa (−74.85%), Brazil (−70.86%), Russia (−69.08%), China (68.2%), and Korea (−31.11%). Even though China has the second lowest average GDP per capita (ppp) in 6 years, its distance to achieve Japan's 2010 level is the second shortest.

Table 3.4 The first five efficiency drivers targeting GDP per capita (ppp) of Japan

Efficiency driver	Brazil	Russia	India	China	Korea	South Africa
Factor #1	Patents	Patents	Literacy rate	Patents	Business R&D	Patents
Factor #2	Higher education enrollment	Business R&D	Internet subscribers	Business R&D	Skilled labor	Internet subscribers
Factor #3	Business R&D	Fair business competition environment	Patents	University–enterprise cooperation	Patents	Skilled labor
Factor #4	Internet subscribers	IP rights protection	Computers in use per capita	Internet subscribers	IP right protection	Business R&D
Factor #5	Basic research	University–enterprise cooperation	Business R&D	Higher education enrollment	Basic Research	Higher education enrollment
Average deficiency percentage of the above 5 drivers	−70.86%	−69.08%	−83.16%	−68.2%	−31.11%	−74.85%

The added value of the above 3D trajectory analysis is (1) to display the NIC paths of the discussed countries on a 48-country landscape; (2) locate the capability region of the four individual capitals for each country; (3) highlight NIC turning points, details of the enhancing or impeding factors, and the dynamics of factor changes for each country; and (4) calculate the distance that each country has to progress in order to achieve the benchmark GDP and identify the first five key factors to reach this goal. This section supplements the previous sections by looking into the contribution of each individual indicator in each year, rather than from the perspective of average scores and average rankings.

In this chapter, we reported our research findings in layman terms with graphs and tables of various types for easy understanding. This study on national intangible assets has uncovered some important findings that are not apparent in other GDP-related studies. For instance, the average distance of the first five efficiency drivers to the same targeted GDP for China is the second shortest, even though its average GDP per capita (ppp) of 2005–2010 is the second lowest (only surpasses India) among the six BRICKS countries. This may explain China's great potential.

Chapter 4
Beyond the 2008 Global Financial Crisis

This chapter reports economic activities after 2010, a timeline regarded as the aftermath of the 2008 global financial crisis, which was formally announced at the end of 2009 (Kehoe 2010; OECD 2010). Although some European countries were still in sovereign debt trouble after 2010 (Greece requested another bailout in early 2012 and Spain asked for external financial assistance in June 2012), we use 2010 as a cutoff point in order to be consistent with the reports of other country clusters in this booklet series. From the statistics introduced in Chap. 3, situations in the six BRICKS countries in 2010 show clear rebounds under the support of each individual economy's stimulus efforts and the gradual revival of global demands.

BRICKS economies began to rebound starting in late 2009 and continued into 2010. The 2008 global financial turmoil did not leave their economies unscathed, yet these six countries have showed good recovery. Strong foreign exchange reserves and growing domestic demand allowed them to withstand the crisis and continue growing (Eghbal 2008). In what follows, we first introduce the general status of the BRICKS countries after 2010 as a whole, and then the condition of each individual country.

The initial notion that emerging economies and countries with strong foreign exchange reserves were immune to the financial crisis has been proven wrong. All six countries were hit hard as foreign investors fled and the demand from the U.S. and Europe dropped sharply. Brazil, China, and Korea experienced weaker exports and India's services sector (accounting for more than half of the GDP in 2007) oriented towards developed economies suffered. Russia was the most vulnerable of all as it is heavily reliant on the hydrocarbon sector (accounting for half of export revenues in 2007), which was severely hit by falling energy prices. South Africa has been very prudent in its financial system and weathered the crisis very well. In addition, all had large foreign exchange reserves at their discretion to increase spending and boost consumer demand, thus making them more resilient to the crisis.

Unexpectedly, this financial crisis brought with it a positive side effect for some countries. It relieved fears of overinflation and made life easier for consumers in countries such as China, India, Korea, and Russia and allowed the governments to

C.Y.-Y. Lin et al., *National Intellectual Capital and the Financial Crisis in Brazil, Russia, India, China, Korea, and South Africa*, SpringerBriefs in Economics 18, DOI 10.1007/978-1-4614-6089-3_4, © The Author(s) 2013

ease interest rates for fueling further growth (Eghbal 2008). In addition, while consumers in developed countries were curbing their spending, BRICKS countries represent a huge growing market with 2.9 billion consumers or 43% of the world's population in 2008.

In what follows, in the sequence of Brazil, Russia, India, China, Korea, and South Africa, we briefly describe what happened in each individual country after 2010.

Brazil

The Brazilian economy began to improve in the second quarter of 2009 (Mendonça 2010). In 2010, its real GDP grew by 7.5%—its fastest pace in 24 years—largely driven by record-low unemployment and a confident new middle class of free-spending consumers (Lopez and Cascione 2011). Consumer and investor confidence revived, boosted by an export recovery mainly from China (Michener 2011; CIA 2011). With credit cards and rising wages, new members of the middle class began buying cars in record numbers, eating out at restaurants, and taking their first-ever foreign vacations (Lopez and Cascione 2011). Brazilian central bank said that the rising capital spending (21.8% for all of 2010) indicates that businesses are confident in Brazil's economic outlook in the coming years (Lopez and Cascione 2011).

Brazil's strong growth and high interest rates made it an attractive destination for foreign investors. Large capital inflows in 2010 contributed to the rapid appreciation of its currency, the real, and led the government to raise taxes on some foreign investments (CIA 2011). As investors forced up the real's value, the currency appreciated by more than 35% since 2005 (Michener 2011). As a result of the strong real, Brazil's manufacturing and industrial exports were hurt and became less competitive both locally and abroad. Another issue of concern was although the tax breaks and loose spending implemented during the crisis helped boost the economy, they also left inflation above 6% in 2010 (Lopez and Cascione 2011). Therefore, controlling inflation becomes one of the most important matters for the Brazilian government in the wake of financial crisis.

Although it was predicted that the robust domestic demand and investment would continue to drive the Brazilian economy in 2011, the domestic market is slowing down because of weaker external demand and market tightening in 2011. According to a Morgan Stanley report, external demand from Europe and the U.S. was still dropping. Furthermore, the withdrawal of policy accommodation and stimulus package in 2011 was partly responsible for the slowdown of emerging market growth. In addition, emerging market economies appeared willing to settle for lower growth rates rather than for a quick boost to growth (Pradhan 2012). In November 2011, OECD reported that, in Brazil, tighter economic policy and weaker external demand have helped cool the economy from the rapid growth rates seen in 2010 with a prospect of suppressing inflation. However, inflationary pressures have not receded and credit growth was still buoyant with activity still expected to grow at below-trend rates over the next 2 years (OECD 2011a).

Russia

Russia's economy began to recover from the financial crisis starting in the second half of 2009. The economic decline bottomed out in the mid-2009 and the economy began to grow in the first quarter of 2010 (CIA 2011). Nearly all of Russia's most important sectors reported strong growth in 2010 (Aris 2011). As mentioned in Chap. 2, Russian government policy response was especially effective in stabilizing the financial sector with a sizeable injection of liquidity into the banking system, which also helped contain unemployment and poverty. According to World Bank Lead Economist for Russia Zelijko Bogetic (2010), the reflow of deposits as early as February 2009 shows that confidence in the Russian banking system was restored. After that, household consumption drove Russia's growth in 2010. In addition, external demand recovery boosted the volume and price of energy resources and other key export commodities (Thessaloniki 2011). Higher oil prices increased Russia's fiscal revenues. In addition, the return of capital inflows began in the last quarter of 2009 and the withdrawal of some stimulus expenditures has lowered the fiscal deficit (Bogetic 2010). Private consumption also recovered on the strength of rising wages and declining unemployment (Thessaloniki 2011).

With the recovery, Russia's situation became stable, which has permitted the central bank of Russia to exit from its previous liquidity support schemes. Thus, the liquidity level of Russian banks has resulted in some of them prepaying their debt in full before maturity (Thessaloniki 2011). Generally speaking, the Russian banking system is well capitalized with a system-wide capital adequacy ratio that stood at 18.4% in the third quarter of 2010. Banks are also highly liquid, with liquid assets as a share of total assets at 25.3% at the end of 2010, higher than pre-crisis rates (Thessaloniki 2011). As a result, profitability has been restored, although credit growth remains weak as credit risks and the banking system's remaining portfolio problems still keep credit constrained (Bogetic 2010).

Outside of banking, industry (including commodity exports) accounted for most of the growth in 2010, growing by an estimated 7.6% with a significant reversal from the −11.6% of 2009; services and construction grew at a more moderate rate of 2.8% (after having declined −5.4% in 2009); yet agriculture contracted by 11.6% in 2010 (against growth of 1.5% in 2009), mainly due to a heat wave and drought in central Russia in 2010 (Thessaloniki 2011). The large decline in agriculture impacted other sectors, such as manufacturing, retail trade, and overall growth adversely, limiting real GDP growth to 3.8% for 2010 (Bogetic 2010).

Bogetic (2010) predicted that the pace of economic growth would moderate in 2011 and would depend on the banking sector's ability to provide long-term credit to facilitate growth in fixed investment. Despite the prospect, critics continue to worry that Russia's economy is overly dependent on the export of oil, gas, and other raw materials. In Russia, other sectors are still not strong enough to pick up the slack from any downturn should global oil and gas prices fall significantly again (Young 2011). In retrospect, the crisis has left Russia's companies in an even more robust shape than they were before (Aris 2011). Over the longer term, the country has confidence to progress towards converging with the developed world.

India

In 2010, the Indian economy rebounded robustly from the global financial crisis. Growth exceeded 8% year-on-year in real terms and merchandise exports, which account for about 15% of GDP, returned to pre-financial crisis levels (CIA 2011). Fortunately, the Indian banking system was not affected much by the global crisis and all financial parameters have remained strong with capital adequacy ratio for the system at 13.65% (tier I ratio at 8.95%) (United Nations 2009). India's ability to come out of the crisis fast was helped by the fact that it is much less dependent than most countries on global flows of trade and capital with the strong support of the country's large and robust internal market, as it relies on external trade for about 20% of its GDP only (the figure for China is roughly double) (Tharoor 2010). In addition, Indians abroad stayed loyal to India and remittances from overseas Indians remained robust, reaching US$46.4 billion in 2008–2009, the bulk of which came from the mainly blue-collar Indian expatriate community in the Gulf countries (Tharoor 2010).

With the worst of the crisis over, in 2010 India became the second most important foreign direct investment (FDI) destination (after China) for 2010–2012 (UNCTAD 2010; Smit 2011). Over the next 2 years India could attract FDI worth US$80 billion, according to Morgan Stanley. During August 2011, exports and imports in US dollars increased by 44.3% and 41.8%, respectively (IBEF 2011). As a result, India entered the club of top 20 exporters of goods and the country also reclaimed its position among top 10 service exporters in 2010 (IBEF 2011).

When compared to countries across the world, India stands out as the fastest growing major economy after China. In order for India's growth to be much more inclusive than what it has been, in 2010–2011 Indian government spending on the social sector was increased to US$30 billion, about US$20 billion was allocated for the development of rural infrastructure, the budget for school education increased by 16% to US$7 billion, and states have access to US$792 million for elementary education. In addition, resource allocation to Ministry of Health & Family Welfare was increased by 25% to US$5 billion (Bajpai 2010a, b). Having these well-planned social advancements in place, the key success factor is the implementation that needs to be strengthened and sustained over time.

With the strong recovery, in 2010 New Delhi reduced subsidies for fuel and fertilizers, sold a small percentage of its shares in some state-owned enterprises, and auctioned off rights to radio bandwidth for 3G telecommunications in part to lower the government's deficit. As of October 2011, foreign institutional investors have invested more than US$7.81 billion in government papers and US$12.95 billion in corporate bonds (Bajpai 2010a, b). However, India's economic growth in 2011 slowed because of persistently high inflation, interest rates, and little progress on economic reforms (CIA 2011).

China

In 2010, the world bounced back from the worst recession since the Great Depression and global growth was once again close to 5%, among which China contributed almost one third (Herd et al. 2011a, b). In the same year, China's GDP was valued at US$5.87 trillion, surpassing Japan's US$5.47 trillion to become the world's second largest economy after the U.S. Predicted by Standard Chartered, China could become the world's largest economy (nominal GDP) by 2020 (Adam 2010). Furthermore, China is now the largest exporter and second largest trader in the world while also being the largest destination for FDI and the sixth largest source of FDI. China's economy was predicted to grow by 10–12% per year between 2010 and 2015 (CIA 2011). In addition to these promising prospects, China announced its 12th Five-Year Plan in March 2011. The government also vows to continue reforming the economy and emphasized the need to increase domestic consumption in order to make the economy less dependent upon exports for GDP growth in the future.

Although impacted by this wave of financial crisis, China did not have a financial sector meltdown, credit crunch, or household debt crisis. As a result, consumers were able to resume spending as soon as confidence returned. The September 2009 issue of the *China Retail Quarterly* suggested not only that consumption is strengthening, but also that consumers outside the marquee cities (Beijing, Shanghai, and Guangzhou) are leading the charge because the masses of rural China are the main source of domestic retail (Fleet 2010). In addition, urban retail sales in the first half of 2009 grew 13.7% year-on-year even in the midst of the global crisis, whereas rural sales grew by 37.6%, according to China's Directorate General of Budget, Accounting and Statistics.

In 2011, China continued to run very large trade surpluses. New construction soared with a government campaign to provide more affordable housing (Bradsher 2011). As the crisis receded, IMF reported that China's ongoing withdrawal of monetary stimulus and the scaling back of public spending was leading to a healthy slowdown. Despite this slightly slower pace, China's growth will continue to propel the world economy forward, contributing one quarter of the overall global growth of 4% (IMF 2011). With the heightened uncertainty in the U.S. and Europe, China's exports still maintained a strong momentum, growing 25% in August 2011 from a year earlier. Import growth was also very strong, led by both commodities and capital goods. Foreign exchange reserves continued to climb, growing by a further US$153 billion in the second quarter of 2011 to US$3.2 trillion. To reduce overreliance on the U.S. and EU markets, China is also increasingly diversifying export destinations toward other emerging markets.

Furthermore, in order to construct a balanced economy that relies more on domestic consumption and directs more resources to non-tradable sectors (such as retail, health, education, and other service industries), China continues to build out its pension system, pursue the comprehensive health-care reform, and increase spending in areas that boost productivity in agriculture and non-tradable sectors.

However, continuous booms and extra cash have sent inflation at the consumer level surging to more than 6%. To fight inflation, in 2011 the Chinese financial authorities set strict administrative quotas on new loans, limiting both banks' ability to lend and businesses' ability to borrow (Bradsher 2011). Combining China's planned slowdown, fighting for inflation, the pressure to appreciate its currency, RMB, from the U.S. and the decreasing demands from advanced countries, the country's huge manufacturing sector is starting to slow and orders are weakening, especially for exports. A question posited to be answered is whether China's slower economic growth and continued inflation would lead to more social tensions, which in turn could damage the economy through weaker investment and more cautious spending (Bradsher 2011).

Korea

The Korean economy was bottoming out from the recession triggered by the global financial crisis, yet the recovery pace was faster than anticipated and has been assessed to be remarkably positive at home and abroad (Kim 2010c). The IMF also gave a positive evaluation of Korea‘s relatively large stimulus package. Since the crisis started, the Korean government has introduced several labor market-related policies, social policies (encompassing roughly a third of the total stimulus package), and reductions in both the corporate income tax and the personal income tax (also encompassing roughly a third of the total stimulus package). The Korean government planned to cut personal income tax rates by 2 percentage points by 2010 and the national corporate income tax rate from 25% (close to the OECD average) to 20% (OECD 2008). In comparison with other OECD countries, the size of Korea's personal income tax reduction was ranked fifth in relation to GDP (Pascha 2010).

The Korean government continued the expansionary fiscal policy in 2010. The stimulus package for 2010 totaled US$148 billion (20.9 trillion won), comprising 37% for public expenditure and 63% for tax reduction. The total amount of fiscal stimulus for 3 years (starting 2008) amounted to US$500 billion (70.5 trillion won) accounting for 6.9% of Korea's GDP (Yoon 2011). Such fiscal stimulus significantly contributed to buffering the negative impact of the crisis in the first half of 2009. Korea's proactive monetary policy response, its export strength, and the international recovery were thus a major force in the recovery (Pascha 2010).

According to *Samsung Economic Research Institute Quarterly*, the Korean economy maintained its growth momentum in the first half of 2010 due to booms in the semiconductor and auto industries. Yet, from July 2010 production in these two backbone industries started to grow at a slower pace than before, resulting in a deceleration in exports and investment. Growth slide in exports was also due to the slowing economic recoveries in developed nations and sluggish growth in emerging economies. In addition, with the gradual scaled back of consumption-boosting government measures, growth in consumption and private spending slowed down as well (Changmock 2010). The pace of growth decelerated from 8.6% to 2.0% between the first and last quarters of 2010; Korean household debt amounted to 124% of household income in 2010 (OECD 2011b). However, the effective exchange rate remained relatively stable over 2010 at about 21% below its early 2008 level, boosting Korea's export competitiveness (OECD 2011b).

After slowing during 2010, growth picked up in early 2011, driven by the acceleration in world trade. The economy regained momentum in the first quarter of 2011, with an output growth of 5.6%. The unexpected March 2011 earthquake and the following nuclear leak disaster in Japan temporarily disrupted Japan's supplies of about a quarter of Korea's imports of parts and capital equipment.

Three years after the financial crisis, the global economy is facing a new crisis of mounting fiscal debt in developed countries, causing low growth of Korean and global economy for a prolonged period. Consequently, the Korean economy is expected to grow 3.6% in 2012, down from 4% in 2011 because the nation's three growth engines—exports, domestic demand and government stimulus—are all likely to lose momentum from 2011 (SERI 2011). Exports, the main stay of Korea's economic growth will lose steam due to the sluggish global economy. Private consumption is expected to remain sluggish in 2012. Slower income growth, high inflation, and an economic slowdown will continue to temper consumer spending.

Up to the end of 2011, resilient recovery in the Korean private sector had yet to surface while the government's policies have largely been exhausted, making it difficult to find a new relief remedy (SERI 2011). Therefore, it becomes clear that both government and companies need to set up a response system for any recurrence of a crisis and reform their structures for sustainable growth in a low-growth economy. In the wake of the 2008 global financial crisis, the major risks for Korea are mainly tied to the strength of the recovery in world trade and its exchange rate. OECD suggested Korea to continue deficit reduction; therefore, significant fiscal consolidation is under way in line with Korea's National Fiscal Management Plan for 2010–2014. Korea's economic outlook is particularly sensitive to demand from China, which accounts for one-third of Korean exports. In addition, to sustain high growth over the medium term requires Korea to continue its structural reforms for enhancing productivity particularly in service (OECD 2011b).

South Africa

Comparing to other developing countries, South Africa's economy has done relatively well (Baxter 2012). After the financial crisis, growth in Africa accelerated with a rate of 2.8% in 2010 and 3.5% in 2011 (Global Finance 2012; Gurria 2010). As external conditions became more favorable, housing prices picked up and leading indicators signaled growing strength. Furthermore, the economy was boosted by the hosting of the 2010 World Cup (Gurria 2010).

A major factor of South Africa's resilience to this global financial crisis was its free floating of the exchange rate, which allowed the rand (South African currency) to follow its function as an adjustment parameter (Draper et al. 2009). In the past, South Africa had not been directly hit by any of the most famous global financial crises, mainly because of its prudent regulatory steps (Dlamini 2010; Draper et al. 2009). A very strong regulatory framework for the financial system, which was supported by strong regulatory institutions, cushioned South Africa from the negative effects of the crisis (Dlamini 2010). In the period preceding the financial crisis, South Africa's banking system also remained stable, and banks were adequately capitalized. The South African banking sector's capital-adequacy ratio remained above the minimum requirement of 9.5% and reached 13.0% at the end of December 2008 (Dlamini 2010). The minimal integration of the South African financial system into the world financial market also minimized its exposure to the riskiest financial products. Another key factor contributing to South Africa's resilience is its well-praised social partners system. Social partners consisted of representatives drawn from the labor, business, government and social communities, who were also responsible for guiding national development and policy implementation (Dlamini 2010).

During the financial crisis, the National Industrial Policy Framework began to promote diversification of the South African economy, particularly through building manufacturing industry and growing associated exports in order to sustain the current account deficit and address its unemployment crisis (Draper et al. 2009). Also during the financial crisis despite increased lobbying from individual labor and business organizations, South African government refused to issue bailout packages, preferring instead to make long-term interventions focused on growth and development (Dlamini 2010). However, its fiscal stimulus included significant levels of public investment in infrastructure till 2012 and was made available in the national budget. This investment was for use in expanding and improving road and rail networks, public transport and port operations, dams, water and sanitation infrastructure, housing construction, information and communications technology, energy-generation capacity and education, and health infrastructure (Dlamini 2010). These measures have laid a firmer foundation for South Africa's economic development.

En route to national recovery, the government also realized that a growing vibrant private sector is the basis for creating and retaining employment in the economy (Baxter 2012). However, the decline of South African Business Confidence Index highlights an unstable and weakening business environment (Baxter 2012; Mail and Guardian 2012). To attract foreign capital and to promote business and consumer confidence, South Africa needs to have stable and predictable economic policies (Baxter 2012).

Chapter 5
Future Perspectives and Policy Implications

By the middle of 2012, more than 3 years have passed since the outbreak of 2008 global financial crisis. Although some countries in the European Union are still in deep financial troubles, it is time to reflect what happened during the past few years and what can be learned from this crisis for future coping strategies.

In general, recovery in the six BRICKS countries is comparatively faster than the recoveries of most other advanced countries. Yet, no matter whether a country is experiencing a fast or slow recovery, lessons must have been learned along the way through the process of formulating coping strategies or implementing stimulus packages to prevent the economy from collapse. In the wake of the financial crisis, most of the countries continue to carry on national reform for future sustainability. Weathering the financial crisis required coordination among public and private sectors. In other words, the economic crisis provides an ideal opportunity to examine the soundness of a nation's economic system and the effectiveness of national governance. In this chapter, we first present the prospects of these six BRICKS countries, afterwards describe their potential challenges in the future, and then end with some policy implications.

Prospects

The original four BRIC countries accumulated high levels of foreign exchange reserves, measuring upwards to US$1,528 billion in China, US$464 billion in Russia, US$266 billion in India, and US$179 billion in Brazil in 2007. These foreign exchange reserves allowed governments to boost public spending in order to support the economy (Eghbal 2008). In the second summit of the four BRIC countries, leaders expressed their common wish to enhance mutual cooperation, including establishing a more democratic and just multi-polar world based on international law, equal cooperation, mutual respect, and collective decision-making. They also pledged to promote the reformation of international financial institutions

C.Y.-Y. Lin et al., *National Intellectual Capital and the Financial Crisis in Brazil, Russia, India, China, Korea, and South Africa*, SpringerBriefs in Economics 18, DOI 10.1007/978-1-4614-6089-3_5, © The Author(s) 2013

(Wang 2010). When South Africa joining the bloc, the alliance became stronger and covered four continents. Introducing Korea is to provide a role model for the other emerging countries for future development.

Brazil

In 2011 as the economic situation in advanced countries was still unsteady, global economy had a prolonged slowdown which spilled over to emerging countries as well. Despite this, Brazil's government pledged to make economic growth and poverty alleviation top priorities, and thus projected good future prospects for Brazil. Export promotion was a main component in its plans to generate growth, taking care to reduce Brazil's vulnerability to international financial market fluctuations at the same time. To increase exports, the government was seeking access to foreign markets through trade negotiations and increased export promotion (TDS 2012). For instance, Brazil and Mexico started negotiations on a free trade agreement (FTA) so that both countries could benefit from the transfer of technology, as well as shared growth and new jobs (Bloomberg 2010). In addition, the EU is Brazil's biggest trading partner and this relationship is likely to grow further as the Brazilian economy expands.

In Brazil, consumer and investor confidence revived and GDP growth turned positive after 2010. Its strong growth and high interest rates make Brazil an attractive destination for foreign investors. With large capital inflows and increasing taxes imposed on the foreign investments, Brazil can formulate long-term national development strategies and implement structural reform for future sustainability.

Another prospect of Brazil is its growing middle class with increasing purchasing power and ongoing large infrastructure projects. As a result, private consumption and domestic investment are likely to become major sources of growth in the years ahead. Its re-industrialization strategy, with high investment in steel, petrochemicals, and defense equipment (including construction of its first atomic submarine), and its continuous technology advancement will provide more job opportunities for years to come (Cárdenas 2008).

Furthermore, Brazil's "Bolsa Família" program created in 2003 seeks to reduce poverty, hunger, and improve social development. Since its creation, the program has expanded rapidly from 3.6 million families in 2003 to 11.1 million in 2008, approximately a quarter of the total population (ILO 2010). Its promotion of school attendance, health, and nutrition also has had positive effect on Brazil's social development. The attempt to break the cycle of intergenerational poverty by improving the well-being and skills of children promotes social equality and is laudable.

For infrastructure, Brazil's hosting of major international athletic competitions every year until the 2016 Rio Olympics will prompt the government to invest in roads, airports, sports facilities, and other infrastructure (TDS 2012). The above stated factors are only examples that may open new prospects for Brazil's future social and economic development. With a dedicated government and proper resource allocation, Brazil's advancing to developed country status is only a matter of time.

Russia

Russia, as the world's largest country, offers a wealth of natural resources, including oil, gas, and strategic minerals. The economy averaged 7% growth since 1998, which resulted in more disposable incomes and the emergence of a middle class (CIA 2011). The burgeoning middle class has led to a growing domestic market in support of future economic development.

The prospect of Russia's development can also be observed from how the government coped with this financial crisis. In early 2009, the Ministry of Economic Trade and Development published an ambitious plan outlining Russian economic goals to 2020. If these goals are reached, Russia will become the largest economy in Europe and the fifth largest in the world, following the USA, China, Japan, and India (CSIS 2012). In the midst of all this, the Russian government also began to pay more attention to the long-term modernization of the economy by instituting a firm anti-corruption campaign and emphasizing augmentation of the judicial system (Palmieri 2012). In addition, the country has been looking into the creative industries as part of its effort to develop and modernize its economy. Emphasis is being given to the fields of design, architecture, and new media. For example, the city of Moscow is undergoing an impressive transformation, merging the old and the new and supporting creative clusters that bring together public and private investors (United Nations 2010). These are all signs of a fair, prosperous, and creative Russian society in the future.

Furthermore, one positive effect of the financial crisis for Russia was that the government decided to implement a number of reforms to shape up the banking sector and to improve its supervisory and regulatory framework. In order to continue consolidation of the banking system, minimum capital requirements were raised to US$3.5 million (Rbl 90 million) in January 2010 and will be raised to US$7 million (Rbl 180 million) in 2012 (Thessaloniki 2011). Banks with insufficient capital will either have to merge with other institutions or else risk being transformed into non-bank credit institutions, or even losing their licenses. Pushing further ahead, the government also decided to partially privatize a number of state-owned banks and to reduce the state's stake in the financial sector. The first deal was completed through the selling of a 10% stake in Bank VTB in February 2011 with more to follow (Thessaloniki 2011). For improving the supervisory and regulatory framework, government plans include tightening the prudential regulation of banks, providing increased powers to the Central Bank of Russia to supervise banks, enforcing counter-cyclical regulatory requirements, restricting foreign currency lending, and differentiating reserve requirements to reduce currency and maturity risks. These measures will help establish a better financial system in preparation for enhancing Russia's international trades.

Another government measure to prevent future financial crises is the investment of the Russian Reserve Fund, capped at 10% of the GDP, in low yield liquid assets, which will be available for use when energy revenues fall and the need to finance a stimulus program arises (Thessaloniki 2011).

Russia is an attractive investment destination with advantages such as a large internal market, quality labor force, extensive resources, and proximity to the EU and other key markets. In addition, Russia's standing in country risk and business environment has improved substantially to 61.5 in 2010, well above the average for the Black Sea Economic Cooperation (BSEC) region of 52.6, although it is still below the EU average of 75.0 (Thessaloniki 2011). Since Russia has made attracting foreign direct investment in coming years, its strategic priority and the development in this area can be anticipated.

Another key prospect for Russia is accession into the World Trade Organization (WTO) to take advantage of reduced tariffs and benefit from trading with member countries. In July 2012, Russia ratified its membership agreement, meaning it became the 156th member of the trade rules club on August 22, 2012 (Reuters 2012). After that, its economic development will begin advancing to another stage (Thessaloniki 2011).

India

India has recovered from the financial crisis faster than most other countries. After combating the crisis, the following features may help the country step up from a Stage 1 (defined by the World Economic Forum) country to a Stage 1.5 or Stage 2 in the near future. They include financial reforms, infrastructure, social safety-net reforms, international trade, and green management.

Even though India's financial sector has proven resilient in the face of the global crisis, the government is still committed to further financial system reforms and improving information access. It has planned for more fiscal consolidation and strengthening fiscal discipline with rules to limit deficit spending (Gurria 2011). In addition, the government encouraged entry of new privately owned banks, which has heightened competition in the sector and yielded efficiency gains (Gurria 2011).

India is determined to tackle "hardware" (ports, roads, airports) and "software" (health care, education) development. As of 2010, the progress has been impressive (Tharoor 2010). Furthermore, the government has outlined measures to boost agricultural growth with matching support from the banks. Indian banks have been consistently meeting targets set for agriculture credit flow, including US$81 billion for the year 2010–2011 (Bajpai 2010a, b).

Over the years, India has built an extensive network of social safety-net programs, including the flagship rural employment guarantee program, which has protected the poor and the returning migrant workers from the extreme impact of recession in the past few years (United Nations 2009). Resources have also been allocated for social sector reform, education, and health and family welfare. These measures may promote a more balanced and fairly shared society.

To boost future international trade, India has strengthened its regional trade agreements (Debroy 2010). For example, India signed an FTA with the Association of Southeast Asian Nations (ASEAN) in August 2009, which went into effect on

January 1, 2010. As a result between 2013 and 2016, tariffs will be eliminated on products including machinery, chemicals, electronics, and textiles (Business in Asia 2012). India and New Zealand also announced the approval for an FTA between the two countries on January 31, 2010 (MFAT 2012). Furthermore, India and Japan reached a bilateral FTA in September 2010. The pact will eliminate import tariffs on most products traded between the two giant Asian economies for 10 years (Masaki 2010).

For environmental protection, US$43 million has been provided for the green revolution through conservation farming, soil health maintenance, water conservation, and preservation of biodiversity (Bajpai 2010a, b).

China

China's ambitious 12th five-year plan, running from 2011 to 2015, builds on decades of unprecedented economic growth. It seeks to transform the economy from a labor-intensive and investment-led powerhouse focused exclusively on GDP growth to more dynamic and technology-based industries (Herd et al. 2011a, b). The target is to build a more sustainable model that balances growth with social harmony and innovation with environmental protection. The government realizes that the key in moving the economy away from labor-intensive industries will be improving the capability to innovate. As a result, the resources devoted to science and technology in China have expanded rapidly in recent years and is now ranked amongst the top countries in total R&D spending and total number of researchers.

The plan also puts great emphasis on generating more equitable growth across the different regions, between rural and urban areas and within urban areas. It acknowledges that increasing domestic demand and improving the income distribution will require a comprehensive development of social services. Whether or not its aspirations can be achieved, the direction in which China's leaders hope to move the country is clear (Herd et al. 2011a, b).

Furthermore, China is shaping its competitive environment in multiple ways. Industries such as education, financial services, health care, and logistics are being deregulated, which further opens the market to foreign companies. Increased attention to food, drug safety, and quality gives companies with a solid reputation for high standards and opportunity. In terms of infrastructure building, China's central government planning includes laying out clear maps for expanding the penetration and capacity of China's rail, power, and communications networks. These expansion plans can ensure that state-owned enterprises and their suppliers (in industries such as construction, equipment, and steel) have secure revenue streams. At the same time, the improved infrastructure will assist in urban-development programs.

Another issue worthy of attention is China's outward FDI. These investments have soared in the past 5 years and even rose very slightly in 2009 when other countries' FDI nearly halved during the financial crisis (Herd et al. 2011a, b). Since China's economic cycle is increasingly influenced by, and induces, fluctuations in the rest of the world, the above stated prospects in China may have some implications for nations and companies that will have associations with China's economic development.

Korea

In the last few years, Korea has proven itself to have a strong and resilient economy that may help it step up to become more advanced in the near future. During the financial crisis, the Korean government sought cooperation with other leading economies, not only to gain concrete access to funds and swap credit lines but also to signal its alignment with important financial players like the USA, Japan, and China (Pascha 2010). Indeed, Korea has become an increasingly active player in the world economy after obtaining OECD membership in 1996. This status brings with it good future prospects for the country.

In terms of national governance, as major developed economies started to retool their systems to achieve fiscal soundness after the breakout of the EU sovereign debt crisis, Korea followed suit (Changmock 2010). That is, in the medium to long term, the Korean government will establish fiscal disciplines to secure fiscal leeway in order to cushion future internal and external economic turmoil.

Environmentally, Korea is strongly committed to promoting green growth through its Five-Year Plan, which envisages spending 2% of its GDP per year through 2013 (OECD 2010). The determination is also evidenced by its announcement of the Green New Deal in January 2009 as part of the financial crisis stimulus package. Developing green technologies will be the side product of this endeavor and will put Korea among the forefront in this area, if the project can be efficiently carried out.

Another positive prospect is that the Korean government announced a program in May 2009 to strengthen its service economy, focusing in many cases on non-tradable. In particular, the plans support education, content industries, IT services, design, consulting, medical services, employment agencies, logistics services, broadcasting, and communication (Pascha 2010). Such a strategy is a manifestation of moving from export-reliant to value-added industrial development.

South Africa

The South African economy benefits from relatively good levels of trade freedom, business freedom, and financial freedom (Heritage 2012). South Africa is the economic hub of sub-Saharan Africa and one of the world's largest producers and exporters of gold and platinum. The following features may forecast South Africa's future prospect for building a sustainable economy.

South Africa has created a regulatory environment that encourages competitiveness and flexibility. The country has managed the transition period after the end of Apartheid rather well (Draper et al. 2009; Heritage 2012) and is in the process of further improving its economic system and governance structure. Unexpectedly, South Africa had the highest inflows in FDI in the crisis year of 2008, at the period of the outflow rush of capital flight from most emerging economies (Draper et al. 2009; Heritage 2012). Apparently, investors still see the macroeconomic conditions in South Africa as favorable to justify a long-term engagement. Furthermore, private investment outpaces public investment indicating the confidence of private investors, which energizing the economy. In addition to relatively good macroeconomic policy and institutions, its microeconomic policies such as competition policies, labor market regulations, regulations in general and trade policies all play a role in the process of improvement.

The financial system of South Africa is stable, and despite the fact that the country was hard hit by the global financial crisis, no exchange rate shock was experienced (Draper et al. 2009; Heritage 2012).

For energy efficiency, there has been a fresh interest in green jobs, the design and implementation of green buildings, and the development of a payment for ecosystem goods and services system, especially among the rural poor (Blignaut 2009). Aggressive adoption of environment protection and products will place South Africa as a forerunner of environment friendly emerging country.

In South Africa, there is a special social partners system among government, labor, business, and community collaboration (Dlamini 2010). The framework has been praised at the international level for bringing together social partners in forging a common response. During the financial crisis, it functioned very well in a sustainable manner. In 1994, the South African government put this structure in place, assuring that social dialogue accompanied the development of economic policies and that civil society had both an opportunity and a platform to articulate its voice in this process.

Challenges

Starting in the second half of 2011, weak economic conditions in the West have been slowing down the global economy. Despite the six BRICKS countries' fast recovery from the financial crisis with proven resilience, they still need to confront challenges that may hamper their future development. This section lists a few, though not an exhaustive list, for readers' reference and exploration.

Brazil

According to the OECD, the Brazilian government spent more, in some respects, in comparison with OECD countries, yet the outcomes were not always commensurate with the level of spending (Gurria 2009). That means there is ample space for enhancing efficiency in government operations. Therefore, investing in strengthening long-term planning, effective implementation with appropriate evaluation criteria and reducing rigidities in the budget should be emphasized. Progress in this area would allow policymakers to identify cost-effective programs to secure adequate funding for these programs and to strategically allocate budgetary resources.

In line with the effective utilization of national resources, another major challenge over the longer term for Brazil is the rise in public spending. General government spending (excluding interest payments) was nearly 32.5% of the gross domestic product in 2008 (Gurria 2009). Deducting the financial crisis stimulus package, this number should still reflect above-average government spending. Mendonça (2010) also warned of the hiking cost of public employment, which has led to a lack of flexibility for public accounts, because the federal government needs to invest around 2% of the GDP on it for a country with a tax revenue of 36% GDP.

Tax reform is also needed in Brazil, as it will encourage businesses to expand and halt predatory industrial policies among the states (Gurria 2009). OECD proposes that the reform can include unifying the different value-added taxes administered by individual states, introducing a similar, single system to replace the various federal levies on company turnover, and reducing taxes and social contributions on wages (Gurria 2009). To reiterate the importance of tax reform, taxes and bureaucratic procedures that used to be taken for granted are now being examined with a more critical eye by businesses, a sector keen to exploit efficiencies (Michener 2011). In addition, corruption and government waste that transfers to higher tax have also come under the close scrutiny. Thus, tax reform has become a big issue in Brazil when its economy relies more on exports and its industries need to have global competitiveness.

Socially, highly unequal income distribution and crime remain pressing problems in Brazil (CIA 2011). Poor quality of education, the worsening of the ports and transport infrastructure, high taxes and high electricity prices also weaken Brazil's competitiveness (Mendonça 2010). Another challenge is to incorporate about one-quarter of the population that lives on less than US$100 per month into more productive and remunerative activities in the labor market (ILO 2010).

Environmentally, the following issues will delay Brazil's progression to a developing country (CIA 2011). They include air and water pollution in Rio de Janeiro, Sao Paulo, and several other large cities; land degradation and water pollution caused by improper mining activities; wetland degradation; and severe oil spills. Deforestation in the Amazon Basin also destroys the habitat and endangers a multitude of plant and animal species indigenous to the area. Furthermore, coping measures are needed to deal with illegal wildlife trade.

In addition to providing a better society for the citizens, Brazil is keen to encourage foreign investment for long-term growth. Therefore, it is essential that the government invests in providing strong, stable infrastructure, and good environment for foreign operations to take roots in Brazil. The above stated challenges are the areas that the Brazilian government can pay attention to for building an attractive context for local as well as foreign institutes to prosper in the future.

Russia

Russia's heavy reliance on single commodity—energy resource exports—made it vulnerable to boom and bust of the global economy (World Bank 2008; CIA 2011). Although the government has embarked on an ambitious program to reduce this dependency and build up the country's high technology sectors since 2007, it has not led to good results (CIA 2011). Now that the crisis is over, challenges remain and require Russian policy makers' attention.

In retrospect, the 2008 recession and the more constrained post-crisis environment has given Russia the opportunity to rethink and accelerate public and financial sector reforms and to promote diversification (Bogetic 2010). Specifically, Bogetic (2010), the World Bank Lead Economist for Russia, provides a list of reform advices as described hereunder. According to suggestions from World Bank, Russia should increase the capacity of its financial sector and ensure the long-term soundness of its financial markets. Bank regulation and supervision as well as operational risk management should be brought up to international standards. Improved disclosure of financial product information and other forms of consumer protection need to be regulated. Further reforms on credit information systems, collateral laws and registers would also help banks increase credit to businesses and households. In general, a new financial framework is needed for a successful financial market transformation.

In addition to the above suggestions, public sector reforms offer the most promise in the short to medium-term (Bogetic 2010). Measures include reducing corruption and arbitrariness in regulatory interventions, increasing transparency, improving the efficiency of public expenditures and large state corporations such as social assistance programs, pension, health, and education system (Aslund et al. 2010; Bogetic 2010; CIA 2011). Furthermore, Russia's public sector employs many more workers than most OECD countries (Dmitriev et al. 2008).

Transiting to the market economy, Russia must also dramatically reduce growth barriers for small- and medium-sized enterprises. Without these companies, Russia's long quest for diversification will not progress forward (Bogetic 2010). For years, smaller or non-energy companies experienced difficulty in accessing capital (CIA 2011) and were put under great pressure during the financial crisis. In 2011, companies other than state-owned enterprises made poor government efficiency an issue, which escalated social tension. As a result, the complaint made the government nervous and consequently it reinvigorated its reform drive (Aris 2011).

Furthermore, the investment climate in Russia should also be improved. In Russia, there is a moderate investment ratio of only 20%, while the Baltic States have a level of 32% (Dmitriev et al. 2008). The old problem of the frequency of tax inspections remains high on the list of private sector complaints; the new problems include the lack of adequate skills among workers and weak transport, electricity and telecommunications infrastructure (Aslund et al. 2010; Bogetic 2010). In other words, large infrastructure investments are needed to facilitate economic development and the application of WTO membership.

In the wake of the financial crisis, Bogetic (2010) predicted that the recovery in Russia will likely provide insufficient jobs; factors include relatively weak growth, skills mismatch, and financing constraints on small- and medium-sized businesses that provide employment opportunities.

If Russia is to achieve its goal of catching up to Western Europe in productivity and income, Russia must find solutions for problems such as its shrinking workforce, underperforming human resources, protection of property rights, and heavy state interference on the private sector. Most of the above stated challenges require strong and substantial reform measures, not just money but also a large, long-term agenda (Bogetic 2010).

India

India weathered the global recession pretty well and fared better than most countries in the global financial crisis of 2008, which was partly due to its insular focus on its domestic market (Arnoldy 2011). But India's rising position fueled more Indian acquisitions abroad and greater foreign investment, which resulted in more exposure to external shock if a global financial crisis strikes again (Arnoldy 2011). In addition, India's fast economic growth has not yet translated into stronger competitiveness internationally, as India ranks only 32 out of 59 countries in IMD's 2011 World Competitiveness ranking (Smit 2011). On the way to prosperity, India may face the following challenges, including structural reforms, social system reform, infrastructure, inflation, education, and environmental issues.

The Indian government needs to step up efforts to restructure public expenditure and reduce the fiscal deficit. For example, between December 2008 and December 2010, India's equity and debt securities nearly doubled from US$91.6 billion to US$171.7 billion (Arnoldy 2011). Relaxing constraints faced by the financial sector and promoting international integration by reducing trade and FDI barriers are also important (Gurria 2011). For financial reforms, special efforts are required to ease wide-ranging and highly prescriptive operating constraints faced by the financial sector for lending, portfolio management, and branch location (Gurria 2011). Structural reform for promoting private investment is essential by easing the impediments of private firms' expansion. In addition, public-sector governance should be made more transparent and accountable.

Social system reforms need to focus on the widespread poverty and corruption, inadequate physical and social infrastructure, low investment in health care, limited nonagricultural employment opportunities, and insufficient access to quality basic and higher education (Bajpai 2010a, b; Gurria 2011; CIA 2011).The social gulf between rich and poor in India is immense. A recent World Bank report found that 42% of the population, or approximately 456 million people, are living below the official poverty line of US$1.25 a day (Ratnayake 2008).

India needs to greatly enhance its infrastructure, especially roads, electricity, and telecommunications to attract foreign investment for economic and national development (Debroy 2010).

Inflation has been an issue of concern in India. Food price inflation has reached critical levels (Debroy 2010). Consumer price inflation and volatile capital flows also need close monitoring (Herd et al. 2011a). Yet, there is little indication in the budget that an effective short-term food inflation management strategy was planned (Bajpai 2010a, b).

For education, teacher effectiveness in the public sector ought to be enhanced through better accountability, incentives, and development pathways. Quality assessment should be strengthened and a higher proportion of funding should be tied to teaching outcomes, such as reducing student drop-out rates and increasing student attendance (Gurria 2011).

Environmental degradation and a lack of effective policies to manage urbanization will also jeopardize India's GDP growth rate (Dobb et al. 2010).

China

Alongside China's rapid growth, there are challenges as well. Recovering from the financial crisis, China faces some immediate economic challenges as described hereunder:

First, during the financial crisis, China's central government allocated a heavy share of the stimulus to the local governments, and some of them became so over-indebted that the central government might be forced to pay the bill after all.

Second, China's lending system favors primarily the big state-owned enterprises. Cash-strapped private small and medium enterprises (SMEs) continue to face poor borrowing conditions, leaving them with little choice other than to turn to the informal financial market (Heilmann and Schmidt 2010).

Third, inflation has long been a problem for China. The largely fixed exchange rate system makes counter-inflationary monetary policies less flexible. In October 2011, Chinese leaders announced that reining in inflation remained the overall priority, dampening expectations that policymakers will attempt to boost growth by loosening credit, lowering interest rates or lifting restrictions in China's residential property market (Pierson 2011).

Fourth, talent shortage is another concern. By 2025 China is expected to have more than triple the number of university graduates it currently produces, many of these degree holders will lack the skills that top Chinese companies and multinationals demand (Devan et al. 2008). With rapid economic development, smaller cities will face intense pressure to attract and create appropriate jobs for an educated workforce. Attracting qualified workers to smaller cities helps diversify the economy and shield smaller cities from future decay similar to what the U.S. Rust Belt experienced during the late twentieth century.

Fifth, aging in China is increasingly becoming an issue, driven largely by low birth rates from the one child policy. Among the four BRIC countries, population percentage under 14 years of age and median age are 26.2%/29.3, 15.2%/38.7, 29.7%/26.2, and 17.6%/35.5 for Brazil, Russia, India, and China, respectively (CIA 2011). That is, both Brazil and India have more population under the age of 14 with median age under 30 than China. The aging problem entails not only a decreasing workforce but also an increasing financial burden for the government.

Sixth, inequality and urbanization also are a point of growing concern. For the next 10 years, urbanization will be a striking dimension of China's rapid ongoing transformation. Continued urbanization will boost incomes, consumption and the size of the service sector, thereby increasing the domestic market and helping with external imbalances. But it calls for further labor market and social policy reforms to facilitate internal migration and reduce inequality. Some of these, notably in the areas of pensions, health care, and the provision of education for migrants' families, need a sizeable financial support. Another inequality is that rural China remains poor even as cities increase in affluence. Accompanying China's rapid economic transformation, inequality has aroused social strife to a certain degree.

Seventh, the environment is one of the China's three most serious challenges in addition to the transition into a truly globalized market economy and political reform (Fleet 2010). Environmental damage—notably air pollution, soil erosion, and the steady fall of the water table—is watched by environmentalists. Other than industrialization, urbanization is a key factor in environmental issues. Harmonious urbanization also raises other challenges, not least on the environmental front. Making the right choices in this respect should help China reach its energy intensity reduction targets (Herd et al. 2011a, b).

Eighth, the Chinese government acknowledges that corruption is endangering political stability and economic growth and has been pursuing an anti-corruption campaign to curb this persistent problem. According to the Business Anti-corruption Portal, China's transition from central planned to market economy not only led to unprecedented economic growth but also created opportunities for public officials and others to enrich themselves illegally.

Korea

For the Korean economy, the past few years can be characterized as a period of a sudden contraction followed by a fast recovery. In the wake of the financial crisis, the challenge for Korea is to learn from the crisis and to conduct necessary reforms for building a stronger economy. There are a myriad of challenges surrounding the Korean economy, some of them are described hereunder.

The main challenge for Korea is to learn from the global crisis on how to stabilize the foreign exchange market (Yoon 2011). Korea has suffered from an exchange rate problem since the global financial crisis erupted. In the second half of 2008, the exchange rate began to increase much faster than the market expected and the Korean won depreciated substantially (Yoon 2011). In particular, the Korean won is highly vulnerable to the movement of foreign funds (Kim 2010a, b).

Another challenge is inflation. The Korean Central Bank Governor said that the greatest challenge facing the Korean economy is inflation and preventing wage increases spurred by inflation expectations. The Bank of Korea tried to stop consumer-price growth exceeding its 4% ceiling for a third straight month in March 2011 (Seo 2011).

Household debt is also expected to remain one of the biggest headaches for policymakers, as snowballing debt and rising interest rates are hurting financial soundness of both households and banks. The country's household debt has been rising at a rate of 12% per year since 2000 to reach US$670 billion (770 trillion won) at the end of September 2010. The ratio reached 153% at the end of 2009, well above that of advanced nations including Canada, the U.S., Japan, and Germany (Kim 2010c). Therefore, on the way to exit from the crisis stimulus plan, a certain degree of international cooperation may be required for Korea's long-term development (Kim 2010c). Learning from other advanced countries and the fast recovery of some emerging economies may be valuable for Korea's future sustainability.

In addition to the money market, the Green New Deal stimulus policy has been criticized on ecological grounds. The opposition parties argue that the river projects may even lead to water shortages in some parts of the country (Pascha 2010). Since this is the long-term project and attached to the proclamation of being a green nation, the issue needs to be carefully studied for building an environment friendly economy.

Korea has also reached a critical wage level with its position endangered by cheaper suppliers elsewhere and more advanced suppliers from the West. Jobless growth and youth unemployment have become an issue, even after the financial crisis. Implementing social policies would ultimately make labor more costly, thus lead to a major upheaval in Korea's competitiveness. Although the Korean government did surprisingly well in finding flexible responses during the financial crisis, the weakness of the SME sector is a persistent issue for the South Korean economy (Pascha 2010). In addition, the SME-related government support addresses a need for structural reform of Korea's current economic system. It was criticized that the support measures for vulnerable SMEs during the crisis do not help to create a competitive economy, but preserve arguably unsustainable business models; in addition, the efficiency and effectiveness of these measures are questionable from a longer term cost-benefit perspective (Pascha 2010).

Global competitiveness is another concern. In order to sustain high growth over the medium term, Korea had to narrow the large labor productivity gap with more advanced OECD economies through reforms, particularly the low productivity in services (OECD 2010). To further strengthen future competitiveness, suggested measures include eliminating domestic entry barriers, accelerating regulatory reform, upgrading competition policy and reducing barriers to trade and inflows of FDI. In addition, it is important to increase labor force participation, notably among women and older persons (OECD 2010), as Korea faces a rapid population aging issue. The share of the elderly in the population is projected to rise from 10% to 14% by 2018, at which point the working-age population will begin to decline. Korea's elderly dependency ratio, now the third lowest in the OECD, is expected to be the fourth highest by 2050. The limited coverage of the public pension system, which has leveled off at about one-third of the working-age population, and the low level and duration of contributions, especially among the self-employed, will challenge the soundness of national well-being (OECD 2008).

In 2007, *Korea Economic Institute of America* published a short report about Korea's economic challenge and mentioned that the appreciation of Korean won against the Japanese yen is a problem. Korea's trade deficit with Japan has continued to widen, as both consumers and manufacturing look to Japan for high tech products. In addition, the evolving economic relationship with China also poses important challenges. China is the largest market for Korea's exports, yet it is also increasingly competing with Korea for markets, as its products are starting to replace Korea's low-tech and medium-low tech goods internationally (Stedman and Cunico 2007). The above stated challenges are the areas that Korean government can contemplate on.

South Africa

South Africa is below the worldwide average with respect to human development, economic freedom, and corruption (Heritage 2012). The most urgent problem is unemployment, which is coupled with weaknesses in the educational system, and the rather low productivity of the South African economy. The following points explain future challenges South Africa may face:

Since 2003, the current account deficit has continuously reached high levels, which raise concerns about South Africa's sustainability (Draper et al. 2009). The financial crisis has highlighted the need for increasing potential growth through a combination of more counter-cyclical macroeconomic policy frameworks. Therefore, structural reforms that lead to higher employment, more competition on product markets, and greater innovation should be implemented (OECD 2010). However, the current cumbersome and restrictive product market regulation hampers the development of entrepreneurship and innovation for long-term growth (Gurria 2010; OECD 2010). In addition, lacking of regulatory transparency also hinders investment (Heritage 2012). According to OECD, politically created barriers to market entry are a very important impediment for South Africa's further development (Draper et al. 2009). The insider dominated wage bargaining has also jeopardized its cost competitiveness. These barriers deter foreign investors and domestic entrepreneurs. Government involvement in the process could help to make the trade-offs between wages, employment, and unemployment clearer to social partners (Gurria 2010).

South Africa's heavy dependence on financial flows into the stock market and the purchase of minerals is likely to erode its industrial base (Fakir 2009). During the financial crisis, several industrial sectors—such as the textile industry—had to be bailed out provides the government a direction for contemplating its future national development strategy (Fakir 2009). To attract investment, the government needs to tackle the bottlenecks in infrastructure, including electricity, transport, and communication (Draper et al. 2009; CIA 2011). A shift towards opening up financial industry to foreign competition would also help upgrade the financial system to support economic growth (Draper et al. 2009).

Despite South Africa's strong macroeconomic policy framework, job creation and productivity growth remain too low to underpin sustained rapid GDP per capita growth. Extreme level of unemployment interacts with inadequate education, poor health outcomes, and crime slowed down the economic development (Gurria 2010). According to its latest Income and Expenditure Survey (2005–2006), 63% of youth between 15 and 30 were unemployed. Of these, 66% (over 1.6 million people) had never before worked. Furthermore, youth constituted 74% of the total number of officially unemployed people in 2005–2006 (OECD 2010). In addition, more than a quarter of all working young people are employed in temporary positions (Frye 2009). Therefore, meaningful job creation, encouragement of entrepreneurship, and wealth creation through a more sustainable growth path of the economy are needed to make full use of South Africa's large supply of underutilized human resources (Fakir 2009; OECD 2010). Also, productivity growth between 2000 and 2005 in South Africa was rather mediocre and slower than in China and India. Therefore, the government should address the problem of low productivity by fostering technological change and basic technologies through enhancing education policy at all levels (Draper et al. 2009).

South Africa's progress in poverty reduction has been slow. Its structural binds continue to trap people in intergenerational poverty. Therefore, it is important to take the crisis rebound initiatives to favor the poor and employ some structural change. Aligning industrial policy, labor market policies, and skills policies to lift the poor and ensure an efficient matching of current and future demands should also be considered (Frye 2009).

South Africa also scores poorly on broad indices of environmental conditions, especially in the area of greenhouse gas emission. As an agricultural country, South Africa has a water supply crisis. The use of surface water for irrigation in agriculture consumes approximately 60% of the total available water annually, with the demand increasing every year (Blignaut 2009). Its economy also heavily relies on carbon-heavy and low-priced coal energy. Furthermore, the country often encountered critical electricity supply shortages. As a result, building an environment friendly economy and solving power supply challenges should be at the heart of South Africa's growth and development policy and better incentives would make the economy more environmentally sustainable (Assubuji et al. 2009; Gurria 2010; OECD 2010). In addition, encourage rural communities to implement renewable energy (biomass and biogas technologies) and other "greening options" is another option. As such, trade can be established in rural communities, and the communities can be involved in contributing economic development, other than restoring the environment (Blignaut 2009).

Policy Implications

Unexpectedly, what started off as a sub-prime mortgage problem in the U.S. would trigger the wide-spreading global financial crisis. Even the emerging economies with strong foreign exchange reserves and large domestic markets are not immune from the impact. This event clearly evidences how interdependent and interconnected the world economies have become.

The causes of this financial crisis are multifaceted. In addition to the much blamed high-risk financial products, undisclosed conflicts of interest, failures of regulation and the credit rating agencies, the sovereignty debt problem broke out in the European Union region adds more complex dimensions to this crisis. Over-spending, over-lending, fragile financial system, and weak national governance surfaced all together at one time. It is a linking system breakdown that failed to withhold the impact of sudden financial shock.

Behind the statistics of GDP growth drop and export volume decrease are the hidden intangibles—the people, systems, governance, market dynamics, and regulation flaws that resulted in this global financial catastrophe. National intellectual capital research reported in this booklet covers human capital, market capital, process capital, and renewal capital encompassing key intangible elements described above. The data analysis describes national intangible assets to a certain degree and provides valuable information for the policy makers to look back the national intellectual capital before the crisis, observe their co-development with the GDP during the crisis, and study its progression afterwards.

By investigating the economic development and national intellectual capital development spanning 2005–2010 at the same time, we have been able to draw the connection between the tangibles and the intangibles for the six BRICKS countries. Some implications can be drawn from our research findings as illustrated hereunder:

First, national intellectual capital development goes together with the economic development and should be regarded as an enhancer of national sustainable growth. The co-development of NIC and GDP with a strong positive correlation (Fig. 3.7) indicates that the higher the national intellectual capital the higher the GDP. Even though countries (except Korea) reported in this volume are at the bottom realm of the continuum, it shows the general development pattern of 48 world major countries. The finding provides these countries a clear direction for national policy planning and for the path to advance from Stage 1 or 2 countries to Stage 3 countries, categorized by World Economic Forum. The reported figures and tables uncover ample information for relevant decision makers to help enhance their national development in a more effective and efficient manner. For example, Table 3.4 shows that among five efficiency drivers, India needs to deal with very basic issues—literacy rate, Internet subscribers, and computers in use per capita; whereas South Africa has to devote more to Internet subscribers, skilled labor, and higher education enrollment. Although both countries are emerging with promising future, the focuses of NIC development are different.

Second, the BRICKS countries are under subtle co-opetition pressure and managing the relationship constructively will lead to a progressing world. A key benefit to form a bloc of emerging countries is to promote trade and leverage each other's resources within the bloc. However, they also compete with each other in the world market with subtle co-opetition (cooperation and competition) pressure. Figures 3.8 to 3.12 show that their NIC-GDP developments are very close to each other. In other words, with their similar stage of development, what they can offer to the world is also similar. Thus, they will compete with each other in an overlapping segment of the market in the increasingly globalized economy. For instance, Brazil relied on China's reviving demand to wade through this financial crisis explains that geographical proximity is no longer an advantage. To avoid future price wars, a better strategy is that each country develops a unique set of competency that complements rather than competes with each other. Currently, China is named the world manufacturer and India a world service provider. Their different national characteristics differentiates their competency. If these countries can all develop complementary capability to help promote global well-being from various perspectives, a better world can be anticipated in the future.

Third, striking a balance between export encouragement and domestic markets expansion will build more resilient economies for BRICKS countries. From our reports, the main reason that these countries were dragged into the 2008 global financial turmoil was the sudden foreign investment withdrawal and sharp drop of export demands. With the experience of this crisis, they are aware of the importance of developing domestic markets to reduce export dependence. Yet, it is easier said than done, as the global demands of the products and services from these six countries are increasing. Exports are still very important as they bring economic development and more advanced technology. However, maintaining the same economic development pattern is no longer sustainable. Therefore, striking a balance between export encouragement and domestic market expansion in their public policies is very important to build a more resilient economy in the future.

Fourth, human capital and renewal capital will be the key intangible assets that decide the forerunners for sustainable national development among emerging countries. In our 48-country intangible–tangible co-development pattern, we found the higher the NIC, the higher the GDP. Although people may argue richer countries have extra money to invest in developing the intangibles, some indicators such as education, computers, mobile phones, and Internet access are basic elements that need to be developed in every country regardless whether they are rich or poor. From Figs. 3.1 to 3.6 Korea is far ahead of the other five countries in human capital, renewal capital, and overall NIC; but not in short-term oriented market capital and process capital. Table 3.1 also shows that Korea is relatively strong in human capital and renewal capital ranking, i.e., the long-term NIC. Furthermore, a comparison of Figs. 3.13 and 3.15 indicates that long-term human capital and renewal capital co-development are confined in a limited scope over the 6 years with no sign of one easily surpassing another, whereas short-term market capital and process capital co-development fluctuates up and down among the countries. That is, advancing long-term NIC takes time and will be the key intangible asset that decides the forerunners of sustainable national development among emerging countries.

Fifth, nations should not be blinded by minor internal economic changes, identifying one's own position on the global landscape and taking proper coping measures are crucial for a nation's sustainable development. National leaders read a variety of statistics. In addition to self-comparing internal progress or regress, assessing their country's relative position in the global arena is also important. We found a radical case in China. Figure 3.23 shows that China's market capital declined by 10 ranks with a small score decrease and its process capital declined by 9 ranks also with a small score decrease from 2005 to 2010. Although China is strong in economic development, particularly during the financial crisis, it still lost some ground in its international NIC standing, a key driver of future national development. On the contrary, Korea had large-scale market capital ranking gains after the financial crisis with a small score increase. These phenomena imply that nations should not be blinded by minor internal development changes, rather identifying their position on the global landscape and taking proper coping measures are crucial in sustaining national development. That is, the name of the game is global standing rather than self-defined improvement.

Sixth, developing intangible national features helps accumulate better market capital and process capital. Short-term NIC contains both tangible and intangible indicators, such as the export of goods and globalization. Although the short-term oriented market capital and process capital are comparatively more susceptible to external ups and downs, investing in the intangible indicators such as globalization and government efficiency (Appendix 3) may help accumulate market capital and process capital in a relatively sustainable way.

Seventh, different from a majority of positive economic statistics about India, NIC analysis reveals warning signals for the country. Most of the economic reports on India are very positive. Indeed, its economic progress is impressive, as the government keeps on implementing structural reforms and important projects are ongoing. Yet, this 2005–2010 NIC-GDP study has picked up some negative signs, especially when compared to other countries in this cluster. For example, India's GDP per capita (ppp) is much lower than China (both have huge populations), total government debt is about three times that of China, its relatively high unemployment rate further increased after the financial crisis (9.62% in 2010), and its consumer price inflation in 2010 is more than double its own 2005 index being the highest among these six countries. The above description discloses that although India recovered from the financial crisis, its strength of recovery is not as good as the other countries. In terms of NIC, India's human capital is the lowest in this cluster, its renewal capital declined over the 6 years, and its best type of capital—market capital—remained virtually constant. These are all warning signs that may not easily surface from objective economic statistics. There is also a warning signal that India's Global Competitive Index (GCI) dropped from 49 in 2005–2006 to 56 in 2011–2012, in spite of world recognition of its economic performance. Global competition is an ongoing game, especially among the BRICS countries (excluding Korea). For sure, India will keep on advancing. Attending to these early warnings will assist India in coming up with better strategies for more effective development. Otherwise, India will only be able to gradually progress toward its goal, while other emerging countries zoom past.

Eighth, Russia should continue with its industry diversification and develop its SMEs. As mentioned previously, Russia is among the most hard hit countries in this financial crisis, mainly due to its single commodity (energy) export. Relying on a single industry export subjects the entire country to global market price fluctuation and increases vulnerability in bad times. The severe negative impact is reflected in its real GDP growth drop to −7.88% in 2009, the lowest in the BRICKS, and its slow rebound to 3.51% in 2010 (Fig. 2.1). Although the Russian government has tried to develop high-tech related industries, its efforts have yet to produce concrete results, very likely because of its limited SMEs. Our NIC analysis reveals that Russia has the highest human capital and renewal capital, yet the lowest market capital in this country cluster, excluding Korea. To capitalize on its strength, Russia can utilize its good quality human resources (human capital) to develop other industries with innovation (renewal capital) for increasing its trading capability (market capital). Now that Russia has adopted the principle of market economy, the government should encourage highly educated and innovative citizens to become entrepreneurs and establish unique industries other than energy for Russia. Therefore, government support for SMEs should be strengthened. With proper support, SMEs are very keen at seeking local as well as foreign alliance, which may help revive the failed high-tech industry development mentioned above.

Ninth, Korea is an ideal benchmark for showing the transformation from an emerging country to a newly developed country. Figures reported in this booklet show that Korea is consistently better than the other five countries with a good distance. In searching for a benchmarking country, aiming too high (such as the Nordic countries) is not realistic for emerging countries. Since Korea has gone through a successful economic transformation only a few decades ago, it should have a lot to offer the other five emerging countries. Korea used to rely on low-tech exports for economic development. Over the years, its products have been successfully advanced to middle-tech and high-tech with the support of its continuous improvement in human capital and renewal capital. Having solid NIC, Korea shows its resilience to this financial crisis and even gains international competitiveness in market capital and renewal capital after the crisis. Korea's transformation is an exemplar for the emerging countries.

Concluding Remark and Emerging Insights

In the wake of the 2008 global financial crisis, there is a lot to reflect upon in both tangibles and intangibles. Although the origins of the 2008 financial crisis are common around the world, the crisis has impacted various economies differently. In advanced economies where it originated, the crisis spread from the financial sector to the real sector. In emerging economies, the transmission of external shocks to domestic vulnerabilities has typically been from the real sector to the financial sector (United Nations 2009). Countries have accordingly responded to the crisis depending on their specific country circumstances, evidenced by how the BRICKS countries have weathered the economic crisis with varying strategies.

Economic developments in the past few years reinforce the observation that economic growth is unequally distributed in the world, which leads to the shift of balance of economic activity. The center of global economic activities will gradually shift to the emerging economies, including countries reported in this booklet. After this financial crisis, the world is undergoing an economic rearrangement that might culminate in a new balance of power, 2011 and beyond. The increasing importance of the most dynamic emerging market economies can be ensured, as they will account for a large part of global growth (Herd et al. 2011a, b; Lopez 2011).

The global recession has also revealed weaknesses in the growth model of these emerging countries, which relies on exports and investments as primary drivers. Although promising, tasks for these countries to accomplish are immense in order for them to really lead future global growth. In hindsight, a new surveillance mechanism is being implemented and important legislative changes are underway internationally. The BRICKS countries also see the need for continuous structural reforms, with macroeconomic, financial, and fiscal management strengthened in an integrated way.

As a recap, to counter the worsening economic conditions worldwide, a total of US$2 trillion in stimulus packages, amounting to approximately 3% of the world gross domestic product, have been allocated, an amount that exceeds the call by the International Monetary Fund (IMF) for fiscal stimulus totaling 2% of global GDP (Nanto 2009). At the beginning of 2011, the World Economic Forum reported that worldwide recovery appeared fairly certain, with economic growth for 2011 and 2012 projected by the IMF at 4.3% and 4.5%, respectively. However in the second half of 2011, due to lasting sovereign debt problems, the European economy tilted back into recession, thereby slowing down the global economy.

This volume briefly reports the main causes of why BRICKS countries were swirled into the financial turmoil, how they came out of it, their positioning in the global NIC landscape, and how the countries utilize their tangible and intangible strength in compensating their weakness for future sustainability. In terms of future NIC development, our other study of the top ten NIC ranking countries reveals that the interplay among the four capitals—human capital, market capital, process capital, and renewal capital—resulted in the best synergic prospect for GDP growth. When their weaker capital (process capital) interacts with stronger capital (market capital), both effects on the GDP are enhanced (Lin 2011). This finding prompts our suggestion for continuous monitoring of NIC development.

At the completion of reporting the impact of the financial crisis on the BRICKS countries from NIC perspective, we have a strong feeling that national success is a capability that accumulates over time meticulously; it does not come within a short burst or from overnight wealth. National economic development is built upon the well-coordinated intangible systems, including human capital, market capital, process capital, and renewal capital systems described in this booklet. As the very origin of the financial crisis—people, systems, governance, market dynamics, and the global interdependence, these intangibles are also the keys to the economic recovery in any place and country.

It is a fundamental function of a government to create and maintain both tangible and intangible infrastructure under which the economic, social, and cultural life of the citizens can prosper. Since it takes time to develop the intangibles, countries that are determined to enhance their so-called soft-power need to consider the added value of national intellectual capital from the start of their policy making.

Appendices

C.Y.-Y. Lin et al., *National Intellectual Capital and the Financial Crisis in Brazil, Russia, India, China, Korea, and South Africa*, SpringerBriefs in Economics 18,
DOI 10.1007/978-1-4614-6089-3, © The Author(s) 2013

Appendix 1 Summary of Main Stimulus Packages of the Six BRICKS Countries

Country	Date	Amount	Item	Sponsor & remarks
Brazil	Sep 2008	#1(reference #) Please see the next column	1. Public bank could purchase the stock of real estate firms and other sectors (US$880 mn to US$1.09 bn) 2. Loans to companies for refinancing their external debts (US$20 bn) 3. On January 21, 2009, the central bank cut its basic rate by 100 basis points (from 13.75% to 12.75%) 4. Expansion of borrowing capacity of PETROBRAS and the National Bank (BNDES) to keep up planned investment levels (US$5.3 bn) 5. US$6.5 bn in support for the agricultural sector 6. A 12% increase in the minimum wage	
		#1 Please see next column	Fiscal policy: 1. Injection of over US$43.8 bn to keep up consumption levels 2. Ensure a budget of at least US$11.4 bn for spending 3. The sectors worst hit by the crisis were granted fresh tax cuts and more resources 4. US$9.2 bn earmarked for the Growth Acceleration Program in the 2009 budget 5. US$6.3 bn reserved for the Sovereign Fund on projects 6. Release US$17.5 bn to help Brazilian firms get raw materials 7. A series of tax cuts totaling around US$3.7 bn to boost consumption 8. Extra US$4.2 bn of government investment, in addition to US$16.6 bn already planned 9. US$2.5 bn Treasury Bond for infrastructure investments	
		#2	Lowering car tax	
		#3 US$221.4 bn	Infrastructure building—focusing on transport system and energy segment.	Central government

	2009	#4	The 2009 stimulus package amounted to a US$20 bn injection into the economy, equivalent to 1.2% of Brazil's GDP, which composed of: 1. Infrastructure investment (41.5%), including My house, My Life housing program for low- and middle-income households 2. Tax cuts (35%) 3. Interest rate subsidies (15.5%) 4. Extraordinary budget transfers to municipalities (5.5%) 5. Extension of the Family Fund (1.5%) 6. Extension of unemployment insurance benefits (1%)	
	March 2009	#4 #5 US$40 bn (BRL72 bn)	Budget for new homes for low- and middle-income families program was increased from US$19 bn to US$40 bn (BRL72 bn).	
	2009	#6	Reduced deposit requirements, bought portfolios from smaller banks, and provide loans to micro- and small enterprises	Central bank
	2008/2009	#6	Public banks provided about 41% of all loans (18.5% of GDP). Reduced the basic interest rate to its lowest level in 20 years (8.75%, equivalent to a real annual rate of less than 5%)	Central bank
	March – 4th quarter, 2009	#6	Temporarily reduced the industrial products tax on automobile sales, household electrical appliances and construction inputs. Lower income tax rates for middle-income families.	Central government
		#6	Extended payment deadlines for various taxes.	The federal government and some states
		#6 US$9 bn (BRL20.5 bn)	Federal government investment rose 12.7%, reaching US$9 bn (BRL20.5 bn)	Federal government #7 From 2008–2010, US$152 bn (around 15% of 2008 GDP)
Russia	Late 2008	#7 US$200 bn (about 7% GDP)	Increase liquidity in the banking sector and aid Russian firms unable to roll over large foreign debts coming due.	Central government

(continued)

Appendix 1 (continued)

Country	Date	Amount	Item	Sponsor & remarks
	Sep 17, 2008	#8 US$44 bn (RUB1.13 tn)	Loan to the three biggest banks, Sberbank, VTB Bank and Gazprombank to boost liquidity	Central government
	Nov 20, 2008	#9	Corporate profit tax rate (24% in 2008) was reduced to 20%	Central government
	Nov 2008	#9 US$20 bn	Surplus package worth US$20 bn was earmarked for cushioning Russian economy	
	2008/2009 2009 Apr 2009 to Mar 2010	#10 US$11 bn (RUB2.9 tn) about 6.7% of GDP	1. Direct support to 295 strategically and 1,148 regionally important firms across various economic sectors (about 75% are large firms) 2. On-the-job training, creating temporary work programs for the unemployed, ensuring direct support to households, facilitating internal labor mobility, and providing support for small and medium enterprises 3. Increase child allowances by 220%, unemployment benefit budget by 70% and 20% of the pensions of the lowest 30% pensioners	Up to about US$8.1 bn (RUB276 bn) from the federal budget #7 From 2008–2010, US$101 bn (8% of 2008 GDP)
		#11	Increased government expenditures from 33.7% of GDP to 40.6% of GDP	Central government
India	Dec 7, 2008 (1st stimulus)	#12 US$4.2 bn (INR200 bn)	Excise duty reduction by 4% in 2008 (extended in the 3rd stimulus) Import Duty on Naphtha for use in the power sector was eliminated (extended in the 3rd stimulus)	Indian Central government #13—1st to 3rd stimulus package amount 1.8% of GDP
	1st stimulus	#13 US$73 mn (INR3.5 bn)	Interest subvention of 2% up to March 2009 Provide guarantees for exports to difficult markets/products Exporters refund of service tax on foreign agent commissions of up to 10% of FOB value of exports	Pre- and post-shipment export credit for labor intensive exports and SME sector Back-up guarantee

1st stimulus	#13 US$229 mn (INR11 bn)	Additional funds to ensure full refund of Terminal Excise duty/ Central Sales Tax (CST)	
1st stimulus	#13 US$833 mn (INR40 bn)	Refinancing the Indian National Housing Bank	
1st stimulus	#13 US$1.5 bn (INR70 bn)	Support incremental lending, either directly to medium, small and micro enterprises, or indirectly via banks	
1st stimulus	#13 US$292 mn (INR14 bn)	Additional allocation to clear the entire backlog in Technology Upgradation Fund (TUF) scheme	For textile industry
1st stimulus	#14	Reduced lending rates on housing loans for low- and middle-income segments (reduced lending rate to 6.5% and borrowing rate to 5%) Removed duty on export of iron ore fines and cut levy on iron lump exports from 15% to 5%.	Indian Central government Reserve Bank of India
Jan 2, 2009 (2nd stimulus)	#13 US$2.1 bn (INR100 bn)	The India Infrastructure Finance Company Limited (IIFCL) refinanced bank lending of longer maturity to eligible infrastructure projects, such as highway	Raise US$2.1 bn (INR100 bn) through tax-free bonds by March 2009
2nd stimulus	#15 US$6.2 bn (INR300 bn) US$5.2 bn (INR250 bn)	The "all-in-cost" ceiling of external commercial borrowing was removed. Increase denominated corporate bonds from US$6 bn to US$15 bn. Provide liquidity support to non-banking finance companies	India's 2nd stimulus package is mainly boosting up what has been planned in the 1st stimulus package.
Feb 24, 2009 (3rd stimulus)	#12	Across-the-board cut on all products other than petroleum and those with rate less than 4%	

(continued)

Appendix 1 (continued)

Country	Date	Amount	Item	Sponsor & remarks
	3rd stimulus	#16	Service Tax Rates reduction by 2%, from 12% to 10% Central Excise Rate reduced by 2%, from 10% to 8% to benefit home appliance and commodity industries	India's 3rd and 4th stimulus packages are mainly based on the first two packages with additions focused on specific sectors
	Feb 26, 2009 (4th stimulus)	#17 US$67.7 mn (INR3.25 bn)	Special Package for Leather and Textiles sector Gem and Jewelry export–import restrictions on worked corals removed	Amounted to 3.5% of GDP for 2008–2009
China	Nov 9, 2008	#18 #19 #20 US$586 bn (CNY4 tn) (19% of 2008 GDP) * CNY is the formal name of Chinese currency, also known as RMB.	More than 86% of the PRC's stimulus package went to infrastructure spending, out of which: 45% was for road, rail, and airport infrastructure (US$ 263.7 bn) (CNY1.8 tn), 9.5% was for improving electricity, water, and road infrastructure in rural areas (US$54.2 bn) (CNY370 bn), 7% was for low-income housing (US$41 bn) (CNY280 bn), and 24.7% was for the reconstruction of towns devastated by the May 12, 2008, earthquake. The remainder of the stimulus package went to health care and education (US$5.9 bn) (CNY40 bn or 1% of the total package), ecological and environmental protection (US$51.3 bn) (CNY350 bn or 8.8%), and technical innovation (US$23.4 bn) (CNY160 bn or 4%).	Central government and local government
	Jan 12, 2009	#19 US$87.8 bn investment spending	Scientific and technical innovation/upgrades	
	Jan 21, 2009	#19 US$124.3 bn	Health-care reform to provide basic medical security for all citizens, improve the quality of medical services, and make medical services more accessible	

Korea	Sep 26, 2008	#18, #21 US$36 bn (KRW50 tn, 5.14% of 2008 GDP) (spread over four years) for Green New Deal	1. Inject US$10 bn into the currency swap market 2. Tax reduction for lower-income taxpayers, income support for low-income households, low-income households received a cash stipend 3. Support for small and medium-sized enterprises and investment. 4. Green New Deal comprises nine major projects including: US$6 bn to improve energy conservation in villages and schools, US$7 bn on mass transit and railroads, and US$11 bn on river restoration. 5. Permanent personal income tax cut (to be executed in 2009 and 2010). 6. The maximum per-person deduction was raised.	More than a quarter of South Korea's stimulus measures consist of tax measures. Social policy measures encompass roughly 1/3 of the total stimulus package and reductions in both the corporate income tax and the personal income tax also account for almost 1/3 of stimulus volume in 2009.
	Oct 19, 2008	#22 #23 #24 US$100 bn US$30 bn	Loan guarantees US$30 bn to banks and exporters in urgent need of foreign exchange	US$30 bn—prop up the Korean banking system
	Oct 27, 2008 Dec 12, 2008	#23 #24 Please see the next column	1. Reduced interest rate from 5% to 4.25% and the interest rate on the Aggregate Ceiling Credit Program from 3.25% to 2.5% to enhance bond market conditions and alleviate credit crunch of SMEs. 2. Expand the currency swap arrangement with the Bank of Japan from US$3 bn to US$20 bn on Dec 12, 2008. 3. Korean government expanded fiscal budget by more than US$34 bn (KRW50 tn) (over 5% of GDP) since the second half of 2008, including: (a) US$3 bn (KRW4.4 tn) through tax refund on oil consumption and others (b) US$3.1 bn (KRW4.6 tn) fiscal spending of US$7.8 bn (KRW11.4 tn) (c) US$6.9 bn (KRW10.2 tn) tax cuts (d) US$11.7 bn (KRW17.2 tn)—revised fiscal budget and supplementary budget in 2009.	Currency swap arrangements with the U.S., Japan, and China, which brought US$90 bn additional liquidity to mobilize funds #24 The expansion in fiscal stimulus package after the second half of 2008 reached US$34 bn (KRW50 tn), which accounted for more than 5% of Korea's nominal GDP in 2008

(continued)

Appendix 1 (continued)

Country	Date	Amount	Item	Sponsor & remarks
	The 2nd half of 2008,	#24 Please see next column	1. Provided US$55 bn of liquidity directly into the foreign exchange market 2. Increased the Foreign Exchange Stabilization Fund by issuing the US$3 bn Global Bond. 3. For 2008 and 2009, Korea's total fiscal stimulus package amounts to US$39 bn (KRW49.6 tn), including US$26.1 bn (KRW33.2 tn) of public expenditure and US$12.9 bn (KRW16.4 tn) in tax reduction. The volume accounted for 4.5% of the GDP in 2009. 4. The stimulus package for 2010 will come up with US$17 bn (KRW20.9 tn) comprising US$6.3 bn (KRW7.7 tn) for public expenditure and US$10.7 bn (KRW13.2 tn) won for tax reduction.	The total amount of fiscal stimulus for three years amounts to US$57.4 bn (KRW70.5 tn) accounting for 6.9% of Korea's GDP. Korean government agreed with ASEAN+3 countries to create a Chiang Mai Initiative reserve pool of US$120 bn to be used in case of crises.
	Jan 6, 2009	#25	Details of the Green New Deal: 1. US$13.7 bn (KRW18 tn)—improve the use of the nation's four major rivers and in construction of dams, reservoirs, and other water management facilities. This was expected to create 280,000 new jobs. 2. US$1.5 bn (KRW2 tn)—invest in small and medium dams to tap new water resources and to help control flooding. This was expected to add 30,000 new jobs. 3. US$8.4 bn (KRW11 tn)—build "green" transportation networks like low-carbon emitting railways, bicycle roads, and other public transport systems. This is projected to create 160,000 new jobs. 4. US$2.3 bn (KRW3 tn)—expand existing forest areas. The public sector would help conserve energy and electricity by constructing 2 mn "green" homes and offices.	

South Africa	Feb 2008	#26 US$69 bn	1. Infrastructural expenditure 2. Provided business entities rewards that increase spending of financial resources	
	Jan 2009	#26 US$14.81 mn	Development Bank of Southern Africa provided Airports Company South Africa with financial assistance	
	Feb 11, 2009	# 27 Please see next column	1. Poverty reduction: US$2.5 bn (ZAR25 bn)—for education and health care, US$1.3 bn (ZAR13 bn)—for social assistance grants, US$404 mn (ZAR4 bn)—for school nutrition program, and US$252.5 mn (ZAR2.5 bn)—to municipalities for basic services. 2. Accelerate employment growth: (a) Additional funding over the medium term—Working for Water and Working on Fire programs, (b) US$101 mn (ZAR1 bn)—the Umsobomvu Youth Fund, (c) US$373.7 mn (ZAR3.7 bn)—low-income housing projects, and (d) US$414.1 mn (ZAR4.1 bn)—the second phase of the expanded public works program. 3. Building capacity to grow: US$79.5 bn (ZAR787 bn) infrastructure investment plans, including (a) US$646.4 mn (ZAR6.4 bn)—public transport, roads, and rail networks, (b) US$414.1 mn (ZAR4.1 bn)—school buildings, clinics, and other provincial infrastructure projects, (c) US$535.3 mn (ZAR5.3 bn)—municipal infrastructure and bulk water systems. 4. Reduce regulatory and microeconomic barriers: (a) US$161.6 mn (ZAR1.6 bn)—industrial development and small enterprise support programs, (b) US$181.8 mn (ZAR1.8 bn)—rural development and small farmer support, (c) US$101 mn (ZAR1 bn)—electricity demand management, together with tax incentives for investment in energy efficient technologies.	Added to the budgets of provinces Government works with business and organized labor to protect work opportunities and accelerate skills development.

(continued)

Appendix 1 (continued)

Country	Date	Amount	Item	Sponsor & remarks
	Apr 2009	#28	1. Supporting a new Special Drawing Rights (SDR) allocation of US$250 bn 2. An additional US$100 bn in lending from the Multilateral Development Banks, 3. Ensuring US$250 bn for trade finance	Tripling the resources available to the IMF, to US$750 bn
	Oct 2011	#29 US$3.2 bn	A competitiveness support package over the next six years to boost industrial development, assist enterprises, and accelerate job creation.	

1. Valadão, M., & Porto, N.G. (2009) *Brazilian Response to International Financial Crisis, the Pre-Salt Discoveries by Petrobras and the New Free Trade Areas in the Cities of Brazil*. 15 Law & Bus. Rev. Am, pp. 673–674.
2. Mendonça de Barros, J.R. (2010) The impact of the international financial crisis on Brazil (ARI), Real Instituto Elcano. Retrieved from http://www.realinstitutoelcano.org/wps/portal/rielcano_eng/Content?WCM_GLOBAL_CONTEXT=/elcano/elcano_in/zonas_in/cooperation+developpment/ari38-2010
3. Economy Watch (June 30, 2010a) Brazil economic stimulus package. Retrieved from http://www.economywatch.com/economic-stimulus-package/brazil.html
4. International Labour Office (ILO) (2010) G20 country briefs: Brazil's response to the crisis. G20 Meeting of Labour and Employment Ministers, Washington, D.C. April 20–21, 2010. Retrieved from http://www.dol.gov/ilab/media/events/G20_ministersmeeting/G20-brazil-brief.pdf
5. Novacich, S.E. (May 17, 2011) Minha Casa Minha Vida development, The Rio Times. Retrieved from http://riotimesonline.com/brazil-news/front-page/minha-casa-minha-vida-development/#
6. Economic Commission for Latin America and the Caribbean (ECLAC) (2009) *Preliminary Overview of the Economies of Latin America and the Caribbean 2009* (pp. 78–80). Retrieved from http://www.eclac.org/publicaciones/xml/3/38063/Brazil.pdf
7. Central Intelligence Agency (CIA) (June 15, 2011) The World Factbook. Retrieved from https://www.cia.gov/library/publications/the-world-factbook/geos/rs.html
8. Nicholson, A. (September 17, 2008) Russia offers Sberbank, VTB, Gazprombank $44 billion, Bloomberg. Retrieved from http://www.bloomberg.com/apps/news?pid=newsarchive&sid=a1sZsKZ32kvU&refer=europe
9. Economy Watch (June 30, 2010b) Russia economic stimulus package. Retrieved from http://www.economywatch.com/economic-stimulus-package/russia.html
10. The World Bank in Russia (March, 2009) Russian economic report No. 18. Retrieved from http://siteresources.worldbank.org/INTRUSSIANFEDERATION/Resources/rer18eng.pdf
11. Åslund, A., Bogetic, Z., Sutela, P., & Carothers, T. (May 4, 2010) Russia's response to the financial crisis, Carnegie Endowment. Retrieved from http://carnegieendowment.org/2010/05/04/russia-s-response-to-financial-crisis/2kc
12. Joseph, M. (2009) Global financial crisis: How was India impacted? Prepared for presentation to InWEnt-DIE Conference on Global Financial Governance—Challenges and Regional Responses, September 3–4, 2009 in Berlin, Germany. Retrieved from http://www.die-gdi.de/CMS-Homepage/openwebcms3.nsf/(ynDK_contentByKey)/MSIN-7SVEYB/$FILE/Global%20Financial%20Crisis%20and%20Impact%20on%20India_Berlin030909.pdf

13. Government of India, Ministry of Finance (December 7, 2008) First stimulus package, in *Different Economic Stimulus Packages for Easing Difficulties of Micro, Small and Medium Sector Enterprises* (pp. 1–4). Office of the Commissioner of Industries, Government of Rajasthan. Retrieved from http://rajind.rajasthan.gov.in/Stimulus_Packages.pdf

14. Kumar, R., & A. Soumya. (September, 2010) Fiscal Policy Issues for India after the Global Financial Crisis (2008–2010). *ADBI Working Paper 249*. Tokyo: Asian Development Bank Institute. Retrieved from http://www.adbi.org/files/2010.09.17.wp249.fiscal.policy.issues.india.after.gfc.pdf

15. Government of India, Ministry of Finance (January 2, 2009) Second stimulus package, in *Different Economic Stimulus Packages for Easing Difficulties of Micro, Small and Medium Sector Enterprises* (pp. 5–9). Office of the Commissioner of Industries, Government of Rajasthan. Retrieved from http://rajind.rajasthan.gov.in/Stimulus_Packages.pdf

16. Government of India, Ministry of Finance (February 24, 2009) Third stimulus package, in *Different Economic Stimulus Packages for Easing Difficulties of Micro, Small and Medium Sector Enterprises* (pp. 10). Office of the Commissioner of Industries, Government of Rajasthan. Retrieved from http://rajind.rajasthan.gov.in/Stimulus_Packages.pdf

17. Government of India, Ministry of Finance (February 26, 2009) Fourth set of stimulus provided through foreign trade policy, in *Different Economic Stimulus Packages for Easing Difficulties of Micro, Small and Medium Sector Enterprises* (pp. 11–12). Office of the Commissioner of Industries, Government of Rajasthan. Retrieved from http://rajind.rajasthan.gov.in/Stimulus_Packages.pdf

18. OECD (June, 2009) *Policy Responses to the Economic Crisis: Investing in innovation for long-term growth*. Retrieved from http://www.oecd.org/sti/interneteconomy/42983414.pdf

19. Abidin, M.Z. (July, 2010) Fiscal policy coordination in Asia: East Asian infrastructure investment fund. *Asian Development Bank Institute Working Paper 232*. Retrieved from http://www.adbi.org/working-paper/2010/07/30/3973.fiscal.policy.coordination.asia/east.asian.response.measures/

20. Heilmann, S., & Schmidt, D. (2010) China country report, in Bertelsmann Stiftung (ed.), *Managing the Crisis. A comparative assessment of economic governance in 14 economies*. Gütersloh: Bertelsmann Stiftung. Retrieved from http://www.bertelsmann-transformation-index.de/pdf/China.pdf

21. Pascha, W. (2010) South Korea country report, in Bertelsmann Stiftung (ed.), *Managing the Crisis. A comparative assessment of economic governance in 14 economies*. Gütersloh: Bertelsmann Stiftung.

22. Fackler, M. (October 24, 2008). South Koreans relieving nightmare of last financial crisis, The New York Times. Retrieved from http://www.nytimes.com/2008/10/24/business/worldbusiness/24iht-24won.17217405.html

23. Kim, H.W. (2010a) Macroeconomic policies of Korea to cope with the crisis. EWC / KDI Conference, Honolulu, Hawaii, 19–20 August 2010. Retrieved from http://www.ssc.wisc.edu/~mchinn/08Kim-Aug2010.pdf.

24. Yoon, D.R. (2011) The Korean economic adjustment to the world financial crisis. *Asian Economic Papers* (10.1) 106–127. Retrieved from http://www.mitpressjournals.org/doi/pdf/10.1162/ASEP_a_00058

25. Instaforex (January 6, 2009) South Korea plans green stimulus package. Retrieved from http://instaforex.com/az/forex-news/163.html

26. Economy Watch (June 30, 2010c) South Africa economic stimulus package. Retrieved from http://www.economywatch.com/economic-stimulus-package/south-africa.html

27. Manuel, T.A. (February 11, 2009) *2009 Budget Speech*. Retrieved from http://www.treasury.gov.za/documents/national%20budget/2009/speech/speech.pdf

28. Dlamini, T. (2010) South Africa country report, in Bertelsmann Stiftung (ed.), *Managing the Crisis. A comparative assessment of economic governance in 14 economies*. Gütersloh: Bertelsmann Stiftung.

29. Gerardy, J. (October 26, 2011) South Africa to launch $3.2 bn stimulus package, Africa Review. Retrieved from http://www.africareview.com/Business+++Finance/-/979184/1261984/-/115wckyz/-/index.html

Appendix 2 Important Meetings Held by World Leaders to Address the 2008 Global Financial Crisis

Date	Place	Participants
June 18–19, 2012	Los Cabos, Mexico	G20
June 17, 2012	Los Cabos, Mexico	G7 finance ministers
March 1–2, 2012	Brussels	European Council
January 30, 2012	Brussels	Informal meeting of members of the European Council
December 8–9, 2011	Brussels	European Council
November 12–13, 2011	Honolulu, USA	19th APEC Economic Leaders' Meeting
November 3–4, 2011	Cannes, France	G20
October 23, 2011	Brussels	European Council
September 10, 2011	Marseille	G8 Deauville Partnership finance ministers
September 9–10, 2011	Marseille	G7 finance ministers and central bank governors
July 21, 2011	Brussels	Meeting of heads of state or government of the Euro area
June 23–24, 2011	Brussels	European Council
March 24–25, 2011	Brussels	European Council
March 17, 2011	Japan	G7 finance ministers telephone conference
March 11, 2011	Brussels	Extraordinary European Council
February 4, 2011	Brussels	European Council
December 16–17, 2010	Brussels	European Council
November 21, 2010	Ireland	G7 finance ministers
November 13–14, 2010	Yokohama, Japan	18th APEC Economic Leaders' Meeting
November 11–12, 2010	Seoul, Korea	G20
October 28–29, 2010	Brussels	European Council
October 8, 2010	Washington DC	G7 finance ministers and central bank governors
September 16, 2010	Brussels	European Council
June 26–27, 2010	Toronto, Canada	G20
June 17, 2010	Brussels	European Council
March 25–26, 2010	Brussels	European Council
February 11, 2010	Brussels	Informal meeting of heads of state or government of the European Union
February 5–6, 2010	Iqaluit, Nunavut, Canada	G7 finance ministers and central bank governors
December 10–11, 2009	Brussels	European Council
November 19, 2009	Brussels	Informal meeting of heads of state or government of the European Union
November 14–15, 2009	Singapore	17th APEC Economic Leaders' Meeting
October 29–30, 2009	Brussels	European Council
October 3, 2009	Istanbul, Turkey	G7 finance ministers and central bank governors
September 24–25, 2009	Pittsburgh, USA	G20

(continued)

Appendix 2 (continued)

Date	Place	Participants
September 17, 2009	Brussels	Informal meeting of heads of state or government of the European Union
June 18–19, 2009	Brussels	European Council
June 12–13, 2009	Lecce, Italy	G8 finance ministers
April 24, 2009	Washington DC	G7 finance ministers and central bank governors
April 2, 2009	G20 London Summit United Kingdom	G20
March 19–20, 2009	Brussels	European Council
February 2009	Rome	G7 finance ministers and central bank governors
December 11–12, 2008	Brussels	European Council
November 2008	G20 APEC Peru	G20
November 22–23, 2008	Lima, Peru	16th APEC Economic Leaders' Meeting
November 14–15, 2008	Washington D.C., USA	G7
October 15–16, 2008	Brussels	European Council
October 10, 2008	Washington, DC	G7 Finance Ministers
September 1, 2008	Brussels	Extraordinary European Council
July 7–9, 2008	Tokyo	34th G8 Summit
June 19–20, 2008	Brussels	European Council
June 13–14, 2008	Osaka	G8 Finance Ministers Meeting
March 13–14, 2008	Brussels	European Council
February 9, 2008	Tokyo	G7 Finance Ministers and Central Bank Governors Meeting

For further information:
http://g20mexico.org/en/g20/previous-leaders-summits
http://www.g8.utoronto.ca/finance/index.htm
http://video.consilium.europa.eu/ecofin/
http://www.european-council.europa.eu/council-meetings/conclusions
http://www.apec.org/

Appendix 3 Indicators in Each Type of Capital

Human capital index	Market capital index
1. Skilled labor*	1. Corporate tax*
2. Employee training*	2. Cross-border venture*
3. Literacy rate	3. Openness of culture*
4. Higher education enrollment	4. Globalization*
5. Pupil-teacher ratio	5. Transparency*
6. Internet subscribers	6. Image of country*
7. Public expenditure on education	7. Exports of goods
Process capital index	**Renewal capital index**
1. Business competition environment*	1. Business R&D spending
2. Government efficiency*	2. Basic research*
3. Intellectual property rights protection*	3. R&D spending/GDP
4. Capital availability*	4. R&D researchers
5. Computers in use per capita	5. Cooperation between universities and enterprises*
6. Convenience of establishing new firms*	6. Scientific articles
7. Mobile phone subscribers	7. Patents per capita (USPTO + EPO)

Remarks:

1. Financial capital is the logarithm of GDP per capita adjusted by purchasing power parity.
2. Indicators marked with an asterisk are rated qualitatively using a scale of 1–10.
3. Indicators with objective numbers are transformed to 1–10 scores, by calculating the ratio to the highest number in that indicator and then multiply 10.
4. Readers can refer to the book *National Intellectual Capital: A Comparison of 40 Countries* for model validation details (www.nic40.org)

Appendix 4 Definition of the 29 Indicators

Indicators	Definition
Human capital	
Skilled labor*	Whether skilled labor is readily available
Employee training*	Whether employee training is a high propriety in companies
Literacy rate	Adult (over 15 years) literacy rate as a percentage of population
Higher education enrollment	Percentage of population that has attained at least tertiary education
Pupil-teacher ratio	Ratio of teaching staff to students
Internet subscribers	Number of Internet users per 1,000 people
Public expenditure on education	Total public expenditure on education (percentage of GDP)
Market capital	
Corporate tax encouragement*	Whether corporate taxes encourage entrepreneurial activity
Cross-border venture*	Whether international transactions can be freely negotiated with foreign partners
Openness to foreign culture*	Whether the national culture is open to foreign culture ideas
Attitudes toward globalization*	Whether attitudes toward globalization are generally positive in a given society
Transparency*	Whether transparency of government policy is satisfactory
Country image*	Whether the image abroad of a given country encourages business development
Exports of goods	Exports of goods (percentage of GDP)
Process capital	
Business competition environment*	Whether competition legislation is efficient in preventing unfair competition
Government efficiency*	Whether government bureaucracy hinders business activity
Intellectual property rights protection*	Whether intellectual property rights are adequately enforced
Capital availability*	Whether cost of capital encourages business development
Computers in use per capita	Number of computers per 1,000 people
Convenience of establishing new firms*	Whether creation of firms is supported by legislation
Mobile phone subscribers	Number of subscribers per 1,000 inhabitants
Renewal capital	
Business R&D spending	Business expenditure on R&D (per capita)
Basic research*	Whether basic research enhances long-term economic development
R&D spending/GDP	Total expenditure on R&D (percentage of GDP)
R&D researchers	Total R&D personnel nationwide per capita (Full-time work equivalent per 1,000 people)
Cooperation between universities and enterprises*	Whether knowledge transfer is highly developed between universities and companies
Scientific articles	Scientific articles published by origin of author (per capita)
Patents per capita (USPTO + EPO)	USPTO and EPO total patents granted (per capita)
Financial capital	
GDP per capita (PPP) (from IMF)	Gross domestic product based on purchasing-power-parity (PPP) per capita

* Indicators marked with an asterisk are rated qualitatively using a scale of 1–10

Appendix 5 Forty-Eight Countries by Cluster and by Continent

This booklet series covers a total of 48 countries in 11 country clusters. China, Brazil, and South Africa are grouped twice to maintain cluster completeness. China is in both Greater China region and BRICKS region, Brazil is in both BRICKS region and Latin American region, and South Africa is in both BRICKS region and in Cluster 5.

Eleven country clusters:

1. Portugal, Italy, Greece, Spain
2. Greater China region (China, Hong Kong, Singapore, Taiwan)
3. Brazil, Russia, India, China, Korea, South Africa (BRICKS)
4. Indonesia, Malaysia, the Philippines, Thailand
5. Israel, Jordan, Turkey, South Africa
6. Bulgaria, Czech Republic, Hungary, Romania, Poland
7. Austria, Belgium, the Netherlands, Switzerland
8. France, Germany, Ireland, United Kingdom
9. Argentina, Brazil, Chile, Colombia, Mexico, Venezuela
10. Australia, Canada, Japan, New Zealand, USA
11. Denmark, Finland, Iceland, Norway, Sweden

Forty-Eight countries by continent:

Europe—24 countries

Nordic Europe—Denmark, Finland, Iceland, Norway, and Sweden

Western Europe—Austria, Belgium, France, Germany, Ireland, Netherlands, Switzerland, and UK.

Southern Europe—Greece, Italy, Portugal, and Spain

Eastern Europe—Russia

East Central Europe—Bulgaria, Czech Republic, Hungary, Poland, Romania, Turkey

Americas—8 countries

North America—Canada, USA

Latin America—Argentina, Brazil, Chile, Colombia, Mexico, Venezuela

Australia—2 countries

Australia and New Zealand

Asia—13 countries

East Asia—China, Hong Kong, Japan, Korea, Taiwan

West Asia—Israel, Jordan

Southeast Asia—Indonesia, Malaysia, Philippines, Singapore, Thailand

South Asia—India

Africa—1 country

South Africa

Appendix 6 National Intellectual Capital Scores and Ranking for 48 Countries (2005–2010)

2005–2010	Human capital		Market capital		Process capital		Renewal capital		Financial capital		NIC	
Mean	6.420		5.409		5.345		3.658		9.115		29.946	
SD	1.217		1.085		1.454		2.056		0.690		5.840	
Country	Score	Ranking	Score	Ranking	Score	Ranking	Score	Ranking	Score	Ranking	Score	Ranking
Argentina	5.370	36	3.340	47	3.038	47	1.413	45	8.755	34	21.916	45
Australia	7.205	15	6.199	12	6.938	9	4.927	16	9.695	11	34.964	13
Austria	7.229	14	6.203	11	6.982	8	5.296	12	9.708	10	35.418	11
Belgium	7.666	8	5.519	23	5.924	21	4.706	17	9.635	14	33.450	19
Brazil	4.757	44	4.589	38	3.076	46	1.753	36	8.481	40	22.656	42
Bulgaria	5.493	34	4.929	31	4.054	36	1.589	41	8.646	37	24.710	35
Canada	7.855	6	6.137	13	6.464	15	5.064	14	9.710	9	35.229	12
Chile	5.414	35	6.663	5	5.163	28	1.710	37	8.784	32	27.734	30
China	4.942	39	5.184	28	3.746	40	2.099	33	7.967	44	23.938	40
Colombia	4.941	40	4.677	36	3.951	37	1.435	44	8.348	42	23.352	41
Czech Republic	5.995	30	5.720	18	5.378	25	2.905	24	9.271	27	29.269	25
Denmark	8.545	1	6.662	6	7.642	1	6.587	6	9.665	13	39.101	3
Finland	7.601	9	6.232	10	7.399	3	7.755	3	9.606	17	38.593	4
France	6.956	20	4.141	45	5.730	22	4.500	20	9.585	18	30.912	23
Germany	6.761	23	5.630	19	6.489	14	5.796	9	9.612	16	34.287	17
Greece	5.986	31	4.216	43	4.113	35	2.168	32	9.440	23	25.923	31
Hong Kong	6.564	25	8.090	2	7.209	5	3.277	23	9.794	4	34.934	14
Hungary	6.674	24	4.836	32	5.095	29	2.301	29	9.046	29	27.952	29
Iceland	8.357	3	5.614	21	6.561	12	5.931	8	9.713	8	36.154	8
India	3.926	48	5.620	20	3.556	42	1.954	34	7.285	48	22.340	44
Indonesia	4.062	47	4.832	33	3.257	44	1.519	42	7.594	46	21.263	47
Ireland	7.145	16	6.816	3	6.711	11	4.101	21	9.768	5	34.541	16

(continued)

Appendix 6 (continued)

2005–2010	Human capital		Market capital		Process capital		Renewal capital		Financial capital		NIC	
Israel	8.040	4	5.954	16	6.454	17	7.091	4	9.414	24	36.953	6
Italy	6.471	28	4.155	44	5.184	27	2.745	26	9.479	22	28.035	28
Japan	7.581	10	4.820	34	6.209	18	6.473	7	9.576	19	34.659	15
Jordan	5.345	37	5.341	26	4.411	32	2.212	30	7.865	45	25.174	33
Korea	6.964	19	5.123	29	5.282	26	4.595	19	9.383	25	31.348	22
Malaysia	6.538	26	6.410	9	5.567	23	2.186	31	8.740	35	29.442	24
Mexico	4.871	42	4.491	39	3.226	45	1.237	48	8.781	33	22.606	43
Netherlands	7.312	12	6.761	4	6.879	10	5.163	13	9.739	7	35.855	10
New Zealand	7.123	18	5.999	15	6.460	16	3.701	22	9.380	26	32.662	21
Norway	7.936	5	6.055	14	7.008	7	4.961	15	9.987	1	35.947	9
Philippines	4.772	43	4.464	40	3.317	43	1.268	47	7.481	47	21.302	46
Poland	6.393	29	4.246	41	3.895	38	1.819	35	8.945	30	25.299	32
Portugal	6.882	22	4.801	35	5.470	24	2.467	27	9.224	28	28.844	26
Romania	5.829	32	4.219	42	3.806	39	1.685	38	8.594	39	24.133	39
Russia	5.530	33	3.860	46	3.588	41	2.349	28	8.822	31	24.150	38
Singapore	7.144	17	8.306	1	7.506	2	5.416	11	9.961	2	38.334	5
South Africa	4.708	45	4.976	30	4.472	31	1.677	39	8.472	41	24.306	37
Spain	6.524	27	4.649	37	5.033	30	2.787	25	9.484	21	28.476	27
Sweden	8.385	2	6.563	7	7.174	6	7.780	2	9.673	12	39.575	1
Switzerland	7.575	11	6.481	8	7.374	4	7.988	1	9.753	6	39.172	2
Taiwan	7.244	13	5.823	17	6.026	20	5.425	10	9.526	20	34.044	18
Thailand	5.310	38	5.524	22	4.365	33	1.436	43	8.304	43	24.938	34
Turkey	4.703	46	5.227	27	4.167	34	1.666	40	8.690	36	24.453	36
United Kingdom	6.939	21	5.440	24	6.090	19	4.697	18	9.626	15	32.793	20
USA	7.695	7	5.348	25	6.558	13	6.715	5	9.886	3	36.202	7
Venezuela	4.884	41	2.761	48	2.545	48	1.277	46	8.625	38	20.092	48

Appendix 7 Country Profile: Additional Statistics

1. International reserves in millions of US$

Country	2005	2006	2007	2008	2009	2010
Brazil	61,498	59,196	108,817	194,223	189,398	242,561
China	663,190	877,637	1,204,035	1,684,280	1,956,830	2,463,547
India	137,008	145,854	192,398	299,684	242,345	261,393
Korea	205,375	217,271	243,840	264,171	206,267	272,248
Russia	133,651	198,677	330,334	498,890	368,065	423,318
South Africa	14,148	20,607	23,849	30,615	30,376	37,496

Source: Global Finance http://www.gfmag.com/

2. GDP per capita (ppp) in US$

Country	2005	2006	2007	2008	2009	2010
Brazil	8,603	9,166	9,900	10,528	10,453	11,239
China	4,103	4,749	5,554	6,189	6,786	7,519
India	2,153	2,402	2,677	2,862	3,039	3,339
Korea	22,783	24,662	26,597	27,716	27,938	29,836
Russia	11,882	13,325	14,909	16,043	14,927	15,837
South Africa	8,654	9,336	10,049	10,453	10,238	10,498

Source: World Economic Outlook Database http://www.imf.org/external/ns/cs.aspx?id=28

3. GDP (PPP)—share of world total_%

Country	2005	2006	2007	2008	2009	2010
Brazil	2.80	2.76	2.78	2.85	2.86	2.93
China	9.46	10.14	11.00	11.74	12.95	13.61
India	4.29	4.47	4.67	4.83	5.20	5.46
Korea	1.93	1.94	1.93	1.93	1.95	1.97
Russia	2.99	3.08	3.17	3.25	3.03	3.00
South Africa	0.72	0.72	0.72	0.73	0.72	0.71

Source: World Economic Outlook Database http://www.imf.org/external/ns/cs.aspx?id=28

4. Gross domestic expenditure on R&D as a % of GDP

Country	2005	2006	2007	2008	2009	2010
Brazil	0.93	0.83	1.02	1.1	1.09	1.23
China	1.23	1.33	1.42	1.49	1.54	1.7
India	0.84	0.84	0.61	0.75	0.83	0.8
Korea	2.63	2.98	3.22	3.01	3.21	3.36
Russia	1.17	1.07	1.07	1.12	1.04	1.23
South Africa	0.73	0.78	0.92	0.92	0.95	0.92

Source: IMD World Competitiveness Yearbook 2006–2011

Glossary

12th Five-Year Plan	China's ambitious 12th Five-Year Plan, from 2011 to 2015
AIIF	Asia Investment Infrastructure Fund
ASEAN	Association of Southeast Asian countries (including Brunei, Burma, Cambodia, Indonesia, Laos, Malaysia, the Philippines, Singapore, Thailand, and Vietnam), China, and India
Basel II	Basel II is the second of the Basel Accords, which are recommendations on banking laws and regulations issued by the Basel Committee on Banking Supervision. Basel II, initially published in June 2004, was intended to create an international standard for banking regulators to control how much capital banks need to put aside to guard against the types of financial and operational risks banks face, 8% ratio usually recommended by the Basel agreements.
BBC	British Broadcasting Corporation
BSEC	Black Sea Economic Cooperation (BSEC) region
Cap and trade	Cap and trade is an environmental policy tool that delivers results with a mandatory cap on emissions while providing sources flexibility in how they comply.
CIA	The U.S. Central Intelligence Agency
CPI	Consumer price inflation
CSIS	Center for Strategic & International Studies
ECLAC	Economic Commission for Latin America and the Caribbean
EU	European Union
ETS	Emission trading system
FC	Financial capital
FDI	Foreign direct investment
FTA	Free trade agreement
GCI	Global Competitiveness Index (published by World Economic Forum)
GDP	Gross domestic product

C.Y.-Y. Lin et al., *National Intellectual Capital and the Financial Crisis in Brazil, Russia, India, China, Korea, and South Africa*, SpringerBriefs in Economics 18,
DOI 10.1007/978-1-4614-6089-3, © The Author(s) 2013

GDP per capita (ppp)	Gross domestic product per capita adjusted by purchasing power parity
HC	Human capital
IMD	International Institute for Management Development
IMF	International Monetary Fund
IP	Intellectual Property
IT	Information technology
Long-term NIC	Long-term oriented national intellectual capital, represented by human capital and renewal capital
MC	Market capital
MICEX	Moscow Interbank Currency Exchange (index in Russia)
Green New Deal	To mitigate the impact of the financial crisis, Korea government launched nine major projects totaling US$36 billion (spread over four years), including almost US$6 billion to improve energy conservation in villages and schools, US$7 billion on mass transit and railroads, and almost US$11 billion on river restoration
NIC	National intellectual capital
OECD	Organization for Economic Cooperation and Development
PC	Process capital
ppp	Purchasing power parity
R&D	Research and development
RC	Renewal capital
RTS	Russian Trading System cash index
Short-term NIC	Short-term oriented national intellectual capital, represented by market capital and process capital
SMEs	Small- and medium-sized enterprises
Term of trade	Term of trade means what quantity of imports can be purchased through the sale of a fixed quantity of exports
Three time periods	For the convenience of comparing ranking changes, the six years data were separated into three time periods, namely 2005–2006, 2007–2008, and 2009–2010. Two years each represents "before, during, and after" the 2008 global financial crisis. Even though Euro zone is still in sovereign debt trouble as of mid-2012, the time partition is based on the financial crisis progression in the epicenter. The signs of financial trouble appeared in 2007 and the crisis was declared over in the third quarter of 2009.
Tier 1 capital	Tier I capital is core capital, including disclosed reserves and equity capital, which cannot be redeemed at the option of the holder.
UNCTAD	United Nations Conference on Trade and Development
USPTO	United States Patent and Trademark Office
VAT	Value-added tax
WEF	World Economic Forum that ranked Global Competitiveness Index (GCI)
WTO	World Trade Organization

References

Adam S (2010) China to exceed U.S. by 2020, Standard Chartered says, Bloomberg
Aris B (2011) Russia 2011–2014. Johnson's Russia List. Retrieved from http://www.cdi.org/russia/johnson/russia-economy-forecast-feb-447.cfm
Arnoldy B (2011) Why India could feel the global financial aftershocks this time. The Christian Science Monitor. Retrieved from http://www.csmonitor.com/World/Asia-South-Central/2011/0808/Why-India-could-feel-the-global-financial-aftershocks-this-time
Aslund A, Bogetic Z, Sutela P, Carothers T (2010) Russia's response to the financial crisis. Carnegie Endowment for International Peace. Retrieved from http://carnegieendowment.org/2010/05/04/russia-s-response-to-financial-crisis/2kc
Assubuji P, Luckscheiter J, Ben-Zeev K (eds) (2009) The global economic crisis and South Africa. Perspectives, #3.09, Political Analysis and Commentary from Southern Africa. Retrieved from http://www.boell.org.za/downloads/Perspectives_3-09.pdf
Bajpai N (2010a) Global financial crisis, its impact on India and the policy response. Retrieved from http://pdffinder.net/Global-Financial-Crisis,-its-Impact-on-India-and-the-Policy-Response.html
Bajpai N (2010b) Global financial crisis, its impact on India and the policy response. Retrieved from http://aric.adb.org/grs/papers/Bajpai.pdf
Baxter R (2012) The global economic crisis and its impact on South Africa and the country's mining industry. South African Reserve Bank. Retrieved from http://www.resbank.co.za/Lists/News%20and%20Publications/Attachments/51/Roger+Baxter.pdf in May 2012
Blignaut J (2009) South Africa and sustainable development: a reflection. In: Assubuji P, Luckscheiter J, Ben-Zeev K (eds) Perspectives, #3.09. Retrieved from http://www.boell.org.za/downloads/Perspectives_3-09.pdf
Bloomberg Business Week (2010) Mexico, Brazil start trade negotiations. Bloomberg Business week. Retrieved from http://www.businessweek.com/ap/financialnews/D9JCAK1G0.htm
Bogetic Z (2009) Global crisis and its impacts on Russia: refocusing policy on households. Carnegie Endowment for International Peace. Retrieved from http://carnegieendowment.org/2009/04/15/global-crisis-and-its-impacts-on-russia-refocusing-policy-on-households/ytd
Bogetic Z (2010). Russia: reform after the great recession, International economic Bulletin. Carnegie Endowment for International Peace. Retrieved from http://carnegieendowment.org/2010/03/30/russia%2Dreform%2Dafter%2Dgreat%2Drecession/3g1v
Bradsher K (2011) China, driver of world economy, may be slowing. The New York Times. Retrieved from http://www.nytimes.com/2011/09/24/business/global/chinas-economic-engine-shows-signs-of-slowing.html?pagewanted=all
Branigan T (2009) Chinese officials and analysts play down deflation fears as prices fall. Retrieved from http://www.guardian.co.uk/business/2009/mar/10/china-deflation-prices-fall

C.Y.-Y. Lin et al., *National Intellectual Capital and the Financial Crisis in Brazil, Russia, India, China, Korea, and South Africa*, SpringerBriefs in Economics 18,
DOI 10.1007/978-1-4614-6089-3,

Bryant L (2011) Growth expected to slow in world's largest economies, Voice of America News. Retrieved from http://www.voanews.com/english/news/europe/OECD-- Economic-Slowdown-Likely-to-Continue-129458543.html

Business in Asia (2012) India's free trade agreement with ASEAN: India's first major multilateral FTA, Business in Asia. Retrieved from http://www.business-in-asia.com/asia/india_fta.html in July 2012

Cárdenas M (2008) Global financial crisis: is Brazil a bystander? Retrieved from http://www.brookings.edu/opinions/2008/1015_financial_crisis_cardenas.aspx

Central Intelligence Agency (CIA) (2011) The World Factbook. Retrieved from https://www.cia.gov/library/publications/the-world-factbook/geos/rs.html

Central Intelligence Agency (CIA) (2012) The World Factbook. Retrieved from https://www.cia.org

Chandrasekhar CP, Ghosh J (2008) India and the global financial crisis. Retrieved from http://www.thehindubusinessline.in/bline/2008/10/21/stories/2008102150440900.htm

Changmock S (2010). Recent economic trend in Korea. SERI Quarterly. Retrieved from http://periodicals.faqs.org/201010/2165371651.html

CSIS, Center for Strategic & International Studies (2012) Economic change in Russia. Retrieved from http://csis.org/program/economic-change-russia

Dahlman C (2009). Different innovation strategies, different results: Brazil, Russia, India, China and Korea. In: Chandra V, Erocal D, Padoan PC, Braga CP (eds) Innovation and growth: chasing a moving frontier. OECD and International Bank for Reconstruction and Development/The World Bank

Debroy B (2010) India country report. In: Bertelsmann Stiftung (ed) Managing the crisis. A comparative assessment of economic governance in 14 economies. Bertelsmann Stiftung, Gütersloh

Delasantellis J (2008) Paulson placates China, Russia – for now. Asia Times Online. Retrieved from http://www.atimes.com/atimes/Global_Economy/JI10Dj04.html

Devan J, Negri S, Woetzel JR (2008) Meeting the challenges of China's growing cities, Mckinsey Quarterly. Retrieved from https://www.mckinseyquarterly.com/Economic_Studies/Country_Reports/Meeting_the_challenges_of_Chinas_growing_cities_2152

Dlamini T (2010) South Africa country report. In: Bertelsmann Stiftung (ed) Managing the crisis. A comparative assessment of economic governance in 14 economies. Bertelsmann Stiftung, Gütersloh, 2010. Retrieved from http://www.bertelsmann-transformation-index.de/pdf/South_Africa.pdf

Dmitriev M, Aslund A, Kuchins A, Charap S (2008) Russian economic goals to 2020: Dreams or reality? A CSIS project. Retrieved from http://csis.org/images/stories/Russia%20and%20Eurasia/080829_ruseura_sumdmitrievaslund.pdf

Dobb R, Sankhe S, Vittal I (2010) India's high-stakes urban challenge. McKinsey Quarterly

Draper P, Freytag A, Voll S (2009) Global financial crisis, protectionism and current account deficit: South Africa on the brink? Working paper No. 8 Global Financial Markets, University Jena and University Halle. Retrieved from http://pubdb.wiwi.uni-jena.de/pdf/wp_hlj08-2009.pdf

ECLAC, Economic Commission for Latin America and the Caribbean – United Nations (2009) Preliminary overview of the economies of Latin America and the Caribbean 2009 (pp 78–80). Retrieved from http://www.eclac.org/publicaciones/xml/3/38063/Brazil.pdf

Edvinsson L, Malone MS (1997) Intellectual capital: realizing your company's true value by finding its hidden brainpower. Harper Business, New York

Eghbal M (2008) BRIC economies withstand global financial crisis. Retrieved from http://blog.euromonitor.com/2008/11/bric-economies-withstand-global-financial-crisis.html

Fakir S (2009) Caught between a rock and hard place: the state of play for South Africa's beleaguered development state. In: Assubuji P, Luckscheiter J, Ben-Zeev K (ed) Perspectives, #3.09. Retrieved from http://www.boell.org.za/downloads/Perspectives_3-09.pdf

Fleet JD (2010) Can China manage its economy? The Diplomat. Retrieved from http://the-diplomat.com/2010/01/21/can-china-manage-its-economy/

Frye I (2009) Responses and alternatives: South Africa's response to the crisis. In: Assubuji P, Luckscheiter J, Ben-Zeev K (ed) Perspectives, #3.09. Retrieved from http://www.boell.org.za/downloads/Perspectives_3-09.pdf

Global Finance (2012) Data on GDP and economic information: South Africa. Retrieved from http://www.gfmag.com/gdp-data-country-reports/176-south-africa-gdp-country-report.html#axzz1urt0NIn8 in July 2012

Government of India, Ministry of Finance (2008) First stimulus package. In: Different economic stimulus packages for easing difficulties of micro, small and medium sector enterprises (pp 1–4). Office of the Commissioner of Industries, Government of Rajasthan. Retrieved from http://rajind.rajasthan.gov.in/Stimulus_Packages.pdf

Government of Rajasthan, Office of the Commissioner of Industries (2009) Different economic stimulus packages for easing difficulties of micro, small and medium sector enterprises. Retrieved from http://rajind.rajasthan.gov.in/Stimulus_Packages.pdf Government of India, Ministry of Finance (2008.12.07) First stimulus package

Gurria A (2009) Recent reforms strengthening Brazil's resilience to economic crisis. Retrieved from http://www.oecd.org/document/44/0,3746,fr_21571361_44315115_43294124_1_1_1_1,00.html

Gurria A (2010) Economic survey of South Africa 2010. OECD. Retrieved from http://www.oecd.org/document/21/0,3746,en_2649_34571_45637781_1_1_1_1,00.html

Gurria A (2011) Economic survey of India 2011. Retrieved from http://www.oecd.org/document/36/0,3746,en_2649_33733_48079140_1_1_1_1,00.html

Heilmann S, Schmidt D (2010) China country report. In: Bertelsmann Stiftung (ed) Managing the crisis. A comparative assessment of economic governance in 14 economies. Bertelsmann Stiftung, Gütersloh. Retrieved from http://www.bertelsmann-transformation-index.de/pdf/China.pdf

Herd R, Conway P, Hill S, Koen V, Chalaux T (2011a) Can India achieve double-digit growth? OECD Economics Department Working Papers, No. 883

Herd R, Koen V, Noord P (2011b) China's emergence as a market economy: achievements and challenges. OECD contribution to the China Development Forum 20–21 March 2011, Beijing

Heritage Foundation (2012) Index of Economic Freedom. Retrieved from http://www.heritage.org/index/ranking

IMF (2011) China economic outlook. Retrieved from http://www.imf.org/external/country/CHN/rr/2011/092611.pdf

IMF (2012a) Regional economic outlook: Asia and Pacific, Managing spillovers and advancing economic rebalancing. Retrieved from http://www.imf.org/external/pubs/ft/reo/2012/APD/eng/areo0412.htm

IMF (2012b) Regional economic outlook: Sub-Saharan Africa, sustaining growth amid global uncertainty, International Monetary Fund. Retrieved from http://www.imf.org/external/pubs/ft/reo/2012/afr/eng/sreo0412.pdf

Indian Brand Equity Foundation (IBEF) (2011) Indian economy overview. Retrieved from http://www.ibef.org/india/economy/economyoverview.aspx

International Labour Office (ILO) (2010) G20 country briefs: Brazil's response to the crisis. G20 Meeting of Labour and Employment Ministers, Washington, D.C., April 20–21, 2010. Retrieved from http://www.dol.gov/ilab/media/events/G20_ministersmeeting/G20-brazil-brief.pdf

Jha R (2008) India: confronting the global financial crisis. Retrieved from http://www.eastasiaforum.org/2008/11/15/india-confronting-the-global-financial-crisis/

Joseph M (2009a) Global financial crisis: how was India impacted? Prepared for presentation to InWEnt-DIE Conference on Global Financial Governance – Challenges and Regional Responses, September 3–4, 2009 in Berlin, Germany. Retrieved from http://www.die-gdi.de/CMS-Homepage/openwebcms3.nsf/(ynDK_contentByKey)/MSIN-7SVEYB/$FILE/Global%20Financial%20Crisis%20and%20Impact%20on%20India_Berlin030909.pdf

Joseph M (2009b) Global financial crisis: how was India impacted? Presentation to InWEnt-DIE Conference on Global Financial Governance – Challenges and Regional Responses, September 3–4, 2009 in Berlin, Germany, Indian Council for Research on International Economic Relations (ICRIER), New Delhi

Kehoe T (2010) The current financial crisis in Spain: what should we learn from the great depressions of the twentieth century? University of Minnesota and Federal Reserve Bank of Minneapolis, 18th Barcelona Lecture, Barcelona G.S.E. and Banc Sabadell, February 2010

Khemka S, Chou S, Fonseca O (2008) Feeling the pain: how the financial crisis is affecting Brazil, Russia, India and China. Knowledge@Wharton. Retrieved from http://knowledge.wharton.upenn.edu/article.cfm?articleid=2079

Kim HW (2010a) Macroeconomic policies of Korea to cope with the crisis. EWC/KDI Conference, Honolulu, Hawaii, 19–20 August 2010. Retrieved from http://www.ssc.wisc.edu/~mchinn/08Kim-Aug2010.pdf

Kim HW (2010b) Macroeconomic policies of Korea to cope with the crisis. Conference Proceedings, Global economic crisis: impacts, transmission, and recovery, Honolulu, Hawaii, 19–20 August 2010

Kim JY (2010c) 2011: year of economic challenges. Korea Times. Retrieved from http://www.koreatimes.co.kr/www/news/biz/2010/12/123_78986.html

Kumar R, Soumya A (2010a) Fiscal policy issues for India after the global financial crisis (2008–2010). ADBI Working Paper 249. Asian Development Bank Institute, Tokyo. Available: http://www.adbi.org/working-paper/2010/09/17/4075.fiscal.policy.issues.india.after.gfc/

Kumar R, Soumya A (2010b) Fiscal policy issues for India after the global financial crisis (2008–2010). ADBI Working Paper 249. Asian Development Bank Institute, Tokyo. Retrieved from http://www.adbi.org/files/2010.09.17.wp249.fiscal.policy.issues.india.after.gfc.pdf

Lin CYY (2011) Leveraging intellectual capital for future competitiveness in Finland, Presentation at Tekes, Helsinki on November 10, 2011

Lin CYY, Edvinsson L (2011) National intellectual capital: a comparison of 40 countries. Springer, New York

Lopez L (2011) Brazil sees BRICs rising, euro zone default change. Reuters. Retrieved from http://www.reuters.com/article/2011/09/21/brazil-economy-financeministry-idUSS1E78K07T20110921

Lopez L, Cascione S (2011) Brazil economy surges in 2010: growth seen cooling. Reuters. Retrieved from http://www.reuters.com/article/2011/03/03/us-brazil-economy-idUSTRE7222QZ20110303

Mail and Guardian (2012) Business confidence on a steady decline. Retrieved from http://mg.co.za/article/2012-05-03-business-confidence-on-a-steady-decline

Masaki H (2010) Japan, India reach free trade agreement, The Journal of Commerce. Retrieved from http://www.joc.com/government-regulation/japan-india-reach-free-trade-agreement

Mendonça de Barros JR (2010) The impact of the international financial crisis on Brazil (ARI), Real Instituto Elcano. Retrieved from http://www.realinstitutoelcano.org/wps/portal/rielcano_eng/Content?WCM_GLOBAL_CONTEXT=/elcano/elcano_in/zonas_in/cooperation+developpment/ari38-2010

Michener G (2011) Brazil's economic outlook a mixed bag. Latin America Monitor. Retrieved from http://www.csmonitor.com/World/Americas/Latin-America-Monitor/2011/0728/Brazil-s-economic-outlook-a-mixed-bag

Ministry of Foreign Affairs & Trade (MFAT) (2012) New Zealand – India free trade agreement (FTA), New Zealand Ministry of Foreign Affairs & Trade. Retrieved from http://www.mfat.govt.nz/Trade-and-Economic-Relations/2-Trade-Relationships-and-Agreements/India/index.php in July 2012

Morrison WM (2011) China's economic conditions. Congressional Research Service report for Congress, 7-5700. Retrieved from http://www.fas.org/sgp/crs/row/RL33534.pdf

Mu X (2010) South Africa joins BRIC as full member, Xinhua Net. Retrieved from http://news.xinhuanet.com/english2010/china/2010-12/24/c_13662138.htm

Nanto DK (2009) The global financial crisis: analysis and policy implications. CRS Report for Congress, Congressional Research Service 7-5700, RL34742

Nicholson A (2008) Russia offers Sberbank, VTB, Gazprombank US$44 Billion. Bloomberg. Retrieved from http://www.bloomberg.com/apps/news?pid=newsarchive&sid=a1sZsKZ32kvU&refer=europe

Nicholson A (2009) Russian economy exited recession in third quarter. Bloomberg. Retrieved from http://www.bloomberg.com/apps/news?pid=newsarchive&sid=aon.9KAC7AMw

OECD (2008) Economic survey of Korea 2008: priorities for macroeconomic policy. Retrieved from http://www.oecd.org/document/39/0,3746,en_2649_34577_41832743_1_1_1_1,00.html

OECD (2009a) Economic survey of Brazil. Retrieved from http://www.oecd.org/dataoecd/35/18/43247511.pdf

OECD (2009b) Policy responses to the economic crisis: investing in innovation for long-term growth. Retrieved from http://www.oecd.org/sti/interneteconomy/42983414.pdf

OECD (2010) Economic survey of Korea 2010. Retrieved from http://www.oecd.org/document/24/0,3746,en_2649_34569_45393816_1_1_1_1,00.html

OECD (2011a) Brazil – economic forecast summary (November 2011). Retrieved from http://www.oecd.org/document/21/0,3746,en_33873108_36016449_45274709_1_1_1_1,00.html

OECD (2011b) OECD economic outlook No. 89 May 2011, OECD. Retrieved from http://www.keepeek.com/Digital-Asset-Management/oecd/economics/oecd-economic-outlook-volume-2011-issue-1-preliminary-version_eco_outlook-v2011-1-en

OECD (2012a) Brazil: economic forecast summary (May 2012), OECD. Retrieved from http://www.oecd.org/brazil/brazil-economicforecastsummarymay2012.htm

OECD (2012b) Korea: economic forecast summary (May 2012), OECD. Retrieved from http://www.oecd.org/korea/korea-economicforecastsummarymay2012.htm

OECD (2012c) Russian federation: economic forecast summary (May 2012), OECD. Retrieved from http://www.oecd.org/russia/russianfederation-economicforecastsummarymay2012.htm

OECD (2012d) South Africa: economic forecast summary (May 2012), OECD. Retrieved from http://www.oecd.org/southafrica/southafrica-economicforecastsummarymay2012.htm

Padayachee V (2012) Global economic recession: effects and implications for South Africa at a time of political challenges, Claves de la Economia Mundial, University of KwaZulu-Natal, Durban. Retrieved from http://www2.lse.ac.uk/internationalDevelopment/20thAnniversaryConference/ImpactoftheGlobalFC.pdf retrieved in February 2012

Palmieri DA (2012) Russia and the global economic crisis. ICOSA Connection and Collaboration. Retrieved from http://www.theicosamagazine.com/russia-and-the-global in January 2012

Pascha W (2010) South Korea country report. In: Bertelsmann Stiftung (ed) Managing the crisis. A comparative assessment of economic governance in 14 economies. Bertelsmann Stiftung, Gütersloh

Pierson D (2011) Inflation remains a worry in China. Los Angeles Times. Retrieved from http://articles.latimes.com/2011/oct/15/business/la-fi-china-inflation-20111015

Pradhan M (2012) 2012: split down the middle. Morgan Stanley. Retrieved from http://www.morganstanley.com/views/gef/

Ratnayake K (2008) International financial crisis exposes vulnerability of Indian economy. Retrieved from http://www.wsws.org/articles/2008/oct2008/indi-o13.shtml

Reavis C (2009) The global financial crisis of 2008–2009: the role of greed, fear and oligarchs. MIT Sloan Management Case #09-093

Reinhart, C.M., and Rogoff, K. (2009) *This time is different: eight centuries of financial folly.* Princeton University Press, Princeton

Reuters (2012) Russia to become WTO's 156th member on August 22, Reuters. Retrieved from http://news.yahoo.com/russia-become-wtos-156th-member-aug-22-135121653.html

SAinfo (2011) New era as South Africa joins BRICS, South Africa Information. Retrieved from http://www.southafrica.info/global/brics/brics-080411.htm

Schwab K (ed) (2011) The global competitiveness report 2011–2012. World Economic Forum, Geneva

Seo E (2011) South Korea's biggest economic challenge is inflation, Kim says. Bloomberg. Retrieved from http://www.bloomberg.com/news/2011-03-16/south-korea-s-biggest-economic-challenge-is-inflation-kim-says.html

SERI (2011) Economic outlook for 2012, Samsung Economic Research Institute. retrieved from http://www.seriworld.org/10/wldSeasonRptL.html?sort=C&sectno=1&cont_type=C

Singh A, Takeda M, Schiff J, Utsunomiya K (2012) Transcript of the Asia and Pacific economic outlook, International Monetary Fund. Retrieved from http://www.imf.org/external/np/tr/2012/tr042112.htm

Smit W (2011) Are you a Tata or a Birla? A study on India's best brands for the world. Retrieved from http://www.imd.org/research/challenges/TC051-11.cfm/?MRK_CMPG_SOURCE=webletter-sep-11&wt.mc_id=webletter-sep-11&utm_source=DM&utm_medium=em&utm_campaign=webletter-sep-11obr

Smith J (2011) BRIC becomes BRICS: changes on the geopolitical chessboard, Foreign Policy Journal. Retrieved from http://www.foreignpolicyjournal.com/2011/01/21/bric-becomes-brics-changes-on-the-geopolitical-chessboard/2/

Stedman L, Cunico JB (2007) Competitiveness: Korea's latest economic challenge. Korea Economic Institute of America. Retrieved from http://www.keia.org/sites/default/files/publications/december%2007.pdf

Taindian News (2008) India's foreign exchange reserves drop nearly US$8 billion. Retrieved from http://www.thaindian.com/newsportal/business/indias-foreign-exchange-reserves-drop-nearly-8-billion_100106096.html

TDS, Travel Document Systems (2012) Brazil economy. Retrieved from http://www.traveldocs.com/br/economy.htm

Tharoor S (2010) How India survived the financial crisis. Retrieved from http://www.project-syndicate.org/commentary/tharoor23/English

The World Bank in Russia (2009) Russian economic report No. 18. Retrieved from http://siteresources.worldbank.org/INTRUSSIANFEDERATION/Resources/rer18eng.pdf

Thessaloniki (2011) Russia: country strategy 2011–2014. Black Sea Trade and Development Bank. Retrieved from http://www.bstdb.org/about-us/key-documents/Country_Strategy_2011-2014_Russia.pdf

United Nations (2009) Global financial crisis and monetary policy response: experience of India. United Nations economic and social commission for Asia and the Pacific, Regional High-Level Workshop on "Strengthening the response to the global financial crisis in Asia-Pacific", 27–30 July 2009

United Nations (2010) Creative economy report 2010, United Nations. Retrieved from http://unctad.org/en/docs/ditctab20103_en.pdf

Valadão M, Porto NG (2009) Brazilian response to international financial crisis, the pre-salt discoveries by Petrobras and the New Free Trade Areas in the Cities of Brazil. 15 Law & Bus Rev Am, pp 673–674. Retrieved from http://studentorgs.law.smu.edu/getattachment/International-Law-Review-Association/Resources/LBRA-Archive/15-3/SMB308.pdf.aspx

Wang G (2010) Bright prospects for BRIC, Xinhua Net. Retrieved from http://news.xinhuanet.com/english2010/indepth/2010-04/15/c_13252144.htm

World Bank in Russia (2008) Russian economic report No.17. Retrieved from www.worldbank.org.ru

World Trade Organization (WTO) (2012) Member information: China and the WTO, World Trade Organization. Retrieved from http://www.wto.org/english/thewto_e/countries_e/china_e.htm in July 2012

Yoon DR (2011) The Korean economic adjustment to the world financial crisis. Asian Econ Papers 10(1):106–127. Retrieved from http://www.mitpressjournals.org/doi/pdf/10.1162/ASEP_a_00058

Young A (2011) Russian economy 2011. Retrieved from http://siberianlight.net/russian-economy-2011/

Zini M (2009) Update on the impact of the financial crisis on South Africa. World Bank. Retrieved from http://blogs.worldbank.org/africacan/update-on-the-impact-of-the-financial-crisis-on-south-africa

Author Index

C.Y.-Y. Lin et al., *National Intellectual Capital and the Financial Crisis in Brazil, Russia, India, China, Korea, and South Africa*, SpringerBriefs in Economics 18, DOI 10.1007/978-1-4614-6089-3, © The Author(s) 2013

Subject Index

C.Y.-Y. Lin et al., *National Intellectual Capital and the Financial Crisis in Brazil, Russia, India, China, Korea, and South Africa*, SpringerBriefs in Economics 18,
DOI 10.1007/978-1-4614-6089-3, © The Author(s) 2013

R

S

MIX
Papier aus verantwortungsvollen Quellen
Paper from responsible sources
FSC® C105338

If you have any concerns about our products,
you can contact us on
ProductSafety@springernature.com

In case Publisher is established outside the EU,
the EU authorized representative is:
Springer Nature Customer Service Center GmbH
Europaplatz 3, 69115 Heidelberg, Germany

Printed by Libri Plureos GmbH
in Hamburg, Germany